iBauhaus

iBauhaus

The iPhone as the Embodiment
of Bauhaus Ideals and Design

NICHOLAS FOX WEBER

ALFRED A. KNOPF
New York
2020

THIS IS A BORZOI BOOK
PUBLISHED BY ALFRED A. KNOPF

Copyright © 2020 by Nicholas Fox Weber

All rights reserved. Published in the United States by Alfred A. Knopf, a division of Penguin Random House LLC, New York, and distributed in Canada by Penguin Random House Canada Limited, Toronto.

www.aaknopf.com

Knopf, Borzoi Books, and the colophon are registered trademarks of Penguin Random House LLC.

Excerpts from "The Apples of Cézanne: An Essay on the Meaning of Still-Life" by Meyer Schapiro, originally published in *ARTnews* Annual XXXIV (1968), pp. 34–53. Courtesy of Artnews Media, LLC, New York, and the Estate of Meyer Schapiro.

Library of Congress Cataloging-in-Publication Data
Names: Weber, Nicholas Fox, [date] author.
Title: iBauhaus : the iPhone as the embodiment of Bauhaus ideals and
    design / Nicholas Fox Weber.
Description: First edition. | New York : Alfred A. Knopf, 2020. |
    "This is a Borzoi book published by Alfred A. Knopf." |
    Includes index.
Identifiers: LCCN 2019016976 (print) | LCCN 2019017557 (ebook) |
    ISBN 9780525657293 (ebook) | ISBN 9780525657286 (hardcover)
Subjects: LCSH: iPhone (Smartphone) | Product design. | Modernism
    (Aesthetics) | Bauhaus.
Classification: LCC TS171.4 (ebook) | LCC TS171.4 .W423 2020 (print) |
    DDC 658.5/752—dc23
LC record available at https://lccn.loc.gov/2019016976

Jacket design by Tyler Comrie

Manufactured in the United States of America
First Edition

To Charlie and Gretchen Kingsley

Don't think that our most high and noble art is taught or learned in schools or academies. What you discover there will be reworked as soon as you are able to observe forms and colors with love.

–Paul Cézanne to Georges Rouault, 1906

You can go anywhere from anywhere.

–Anni Albers

iBauhaus

Part I

1.

"So let us therefore create a new guild of craftsmen, free of the divisive class pretensions that endeavored to raise a prideful barrier between craftsmen and artists! Let us strive for, conceive, and create the new building of the future that will . . . rise heavenwards from the million hands of craftsmen as a clear symbol of a new belief to come."

This was the goal for which Walter Gropius founded the Bauhaus, a groundbreaking school and laboratory for modernism. The year was 1919. The Bauhaus would soon have workshops to develop new textiles, chairs, tables, flatware, lamps, children's toys, and a range of other everyday objects that would revolutionize the way human beings lived all over the world. Prototypes would be made for objects that could be manufactured on a large scale, all of them designed to ameliorate daily existence. While performing needed tasks, they would add charm through their aesthetic grace. The stuff of life was to be absent ornament. Honest, clean, easily maintained, and visually appealing; it would create a new emotional ease.

Worldwide, humankind was to have materials of impeccable construction. They would be compact, sleek, and able to achieve multiple functions. These revolutionary objects would facilitate unprecedented possibilities for everyday existence. And they would be playful as well as useful. The sense of fun alongside the candid "This is what I am and this is what I do" was to transform the human spirit. Design—capable of miracles, truthful and alluring—was the new religion.

The Bauhaus started in a former arts-and-crafts school in the small, historic German city of Weimar. In 1925, it moved to ample new headquarters that Gropius designed for it in Dessau, a rather isolated industrial city. The building epitomized the handsome, streamlined style the school advocated for design of every sort. After the new right-wing government forced its closure in 1932, it had one last desultory year in a disused telephone factory in Berlin. All in all, it would last only fourteen years. Like all brilliant experiments in new approaches to living, it was both a utopia and a place that struggled to survive.

Much of what is greatest in human existence—the amazing engines that are our bodies, the earth itself—makes possible and enhances our lives without our necessarily being conscious of the details. The Bauhaus school does not rival the miracles of nature, but it was a determined effort to transform the way things look and give birth to objects that make daily existence infinitely easier. It succeeded. The school's impact greatly exceeds its recognition. Most of the world does not know the name "Bauhaus," but the manifestations of its approach to design are everywhere. The achievements of this institution—where likeminded people, of dramatically different backgrounds but a shared utopian spirit, gathered together and invented the new—pervade our world.

When the Bauhaus opened its sparkling new headquarters in Dessau, everything was straightforward and purposeful. The name itself was what "iPhone" would be: a startling invention, its two punchy syllables based on the familiar, yet without precedent.

2.

In 1983, Steven Paul Jobs, the cofounder of Apple Computer, gave a speech at the Aspen Institute. The institute and its annual summer conference were the brainchildren of Walter Paepcke, a successful Chicago-based businessman who, with his wife, Elizabeth, had developed a profound admiration for the Bauhaus.

In 1939, Elizabeth Paepcke—nicknamed "Pussy"—had discovered Aspen, a former mining town now nearly abandoned, when the pipes froze at her country house south of Denver and she needed a place to take her guests skiing. In 1945, she got her husband there. Walter Paepcke invited Walter Gropius to come redesign the Victorian town, where he acquired property mostly by paying overdue taxes. After Gropius said no, but proffered the advice not just to restore the old but also build modern, Paepcke got the Bauhaus-trained architect Herbert Bayer to come.

Together, Walter Paepcke and Herbert Bayer planned an international celebration of the two hundredth birthday of Johann Wolfgang von Goethe, the philosopher, novelist, color theorist, and poet. Goethe's pantheism had been embraced at the Bauhaus. Goethe was a longtime resident of Weimar, which was among the reasons the small city was so apt for the design school.

Two thousand people attended the events held in Aspen in 1949 in honor of Goethe's creative genius. Among them were the pianist Arthur Rubinstein, the novelist and playwright Thornton Wilder, the philosopher José Ortega y Gasset, and the humanitarian doctor Albert Schweitzer. Only four years after the horrors of World War II had come to an end, the celebrants were making the point that nationality did not matter as much as the capacity, whatever a person's parentage was, to contribute beautifully to human progress.

Herbert Bayer had gone to the Weimar Bauhaus as a student in 1921. Two years later, his streamlined lettering and sparkling graphics had helped establish the identity of the great international Bauhaus exhibition that took place in Weimar in 1923.

Herbert Bayer designed the bold and eye-catching catalogue for the first Bauhaus exhibition. Twenty-six years later, after fleeing to America, he would cofound and design the Aspen Institute, where Steve Jobs both learned and promulgated Bauhaus ideas.

Bayer's bold signage marked the entrance of the show with panache. His posters for the show have a freshness and rhythmic asymmetry that prepared you for the flair of the paintings by Kandinsky and Klee presented inside. Bayer's cover for the exhibition catalogue is dashing. The sans serif typeface he invented for it animates each letter so that the S's and B's and A's and the rest march with triumphant energy. The words "STAATLICHES BAUHAUS IN WEIMAR 1919–1923" cover the surface of the perfectly square page, printed so that the syllables alternate between red and blue, the cheerful colors syncopating in front of a matte black background. This is graphic design that energizes you.

Steve Jobs gave his speech at Aspen forty-five years after Bayer had immigrated to the United States. It was the third summer that he had gone to one of these annual conferences that had begun with the celebration of Goethe. The international roster of visionaries who attended gave new life to the Bauhaus ethos. Jobs well knew the work of Aspen's founder and admired other pioneering Bauhaus designs.

Jobs thrilled to the idea of speaking in the handsome white building Bayer had designed for the Aspen Institute. The sparkling surfaces and crisp lines declare the happy marriage of its machine-made modernism and the natural splendor of the mountain setting. No other educational institution excited Jobs as much as the Bauhaus did. He was in his element at this conference exploring new ways of improving human existence.

3.

The exhibition in Weimar for which Bayer's graphics pulsed like neon had been organized posthaste. Gropius had not felt ready for this show. He had consented to it under duress, and put it together far too quickly according to his standards and agenda. He had no choice other than to hurry.

The wish to perpetuate the new is often both hampered and advanced by compromises begotten by necessity, and this was the case here. Gropius had been informed by the authorities that he urgently needed to validate his educational experiment in order to retain the government funding that was its lifeline. It was essential, right away, to show visitors from afar not only the art by the most advanced of the painters teaching at the school, but also the new designs coming out of the Bauhaus workshops. Its supporters would keep financing the Bauhaus only if there was proof that it was realizing a vision sufficient to inspire the rebirth of civilization they deemed essential after the disasters of the First World War. Were the new designs, brazen and audacious in their break from tradition, applicable for widespread use?

Steve Jobs had already known many such moments and would experience more of them: those times when his baby, Apple Computers, threatened to crumble, or the competition was taking the lead, or the bank balance was plummeting toward zero. He was used to challenges and urgency. The Bauhaus's struggle simply to exist, the desperate grasping for a stronghold, resonated with him. For Jobs, the most effective means of surmounting difficulties and emerging triumphant over desperate situations was the same solution that it had been at the Bauhaus: beautiful and effective design enjoyed by a large public.

The Aspen Institute gave Steve Jobs an audience that would respect his courage and inventiveness. He hoped they would help fulfill his urgent needs, which he would lay out candidly. The rare opportunity to address so many imaginative design professionals inspired him. All the attendees wanted to advance the issues of how objects look and function, to progress beyond old approaches and ameliorate everyday life.

This assemblage of creative people in a remote location in the Rocky Mountains had first had to get themselves to Denver and then to make a long journey by car. Today there are small planes that render the shuttling to the mountain village quick and easy. But none of the original Aspen Institute attendees, except for Walter Paepcke himself, had that sort of money. The difficult trip was worth it. These people of similar goals relished the synergy that emerged when they addressed their mutual passion.

They were a new generation of the Bauhaus family: the spiritual descendants of the highly principled designers and artisans who had produced one innovative and useful object after another in the 1920s. Steve Jobs, already a highly successful and sought-after leader in the burgeoning field of high-tech communication, was equally content to have Aspen on his annual travel schedule. It put him with other people designing for society at large who happily took risks, possibly losing money and sacrificing job security in order to try the unprecedented.

4.

Walter and Elizabeth Paepcke's daughter was among a group of supporters of the Harvard University Art Museums who in 2016 came to visit the Josef and Anni Albers Foundation in Connecticut. For over forty years, I have directed that foundation, created by two of the few remaining Bauhaus masters still alive in the 1970s. (Herbert Bayer and Marcel Breuer were among the others.) Antonia Paepcke DuBrul said that at the Albers Foundation she could feel the spirit that had been dear to both of her parents. With crusty opinions about what she did not like in today's trendy art world, and gusto about the values embraced by her mother and father, DuBrul said that both her parents had sought to establish a lively and diverse community in Aspen. Restaurant busboys and corporate executives, philosophers and bricklayers, could discuss ideas from a point of view of shared interests rather than differences.

In Aspen, they could hike to mountain peaks or attend wonderful concerts when they were not engaged in boundary-breaking conversations. Elizabeth Paepcke, came from a family of distinguished intellectuals—her brother was the diplomat Paul Nitze. DuBrul's father, Walter Paepcke, was one of those

The association of Walter Paepcke (left) and Herbert Bayer (right) in the period of optimism following World War II brought modern design, and advocacy for its increased prominence in American life, to Aspen. They shared a vision based directly on Bauhaus values. Form would follow function more than it used to, but there was still a need for style, and for the capacity of the right details to brighten the day: note the perfectly proportioned glimpses of neatly folded white pocket handkerchiefs adorning each man's jacket.

rare highly successful businessmen with profound aesthetic convictions that permeated every aspect of their business activities. The Aspen Institute was their utopia.

Antonia DuBrul loathed the way that Aspen has become a watering hole for billionaires, now no different in many ways from other chic habitats of the megarich. She said that the Albers Foundation, built at low cost in a Connecticut forest as both a haven for intellectual exchange and a means of preserving the art and ideas of Anni and Josef, had the authenticity her parents had wanted in Aspen. DuBrul observed that the best of the extraordinary values of the Bauhaus, vital to her father and mother in their own organization, was in the air of the Albers Foundation.

When I reminded her of the importance of Aspen to Steve Jobs, Antonia DuBrul simply said, "Well, yes, of course! Guts! Simplicity! The willingness to listen to what people say they need, and to sense what they want even if they don't know it themselves. The importance of design where every millimeter counts—that's what we are talking about, not this self-infatuated nonsense that parades itself as art today!"

In Albers territory, Antonia DuBrul felt as if she were with family. Herbert Bayer's connection with Josef and Anni dated back to the early 1920s. When Steve Jobs made his summer pilgrimages to Aspen, the institute still had had Bayer at the helm.

Bayer had come to America five years after the Third Reich forced the closing of the Bauhaus. Hitler's minions had deemed the Bauhaus too "non-German" in both spirit and population, and the Gestapo had padlocked the doors of the Bauhaus's last stronghold. When Bayer and Paepcke created their haven near a Colorado mining town down at its heels, they intended it to perpetuate the values the Nazis had tried to destroy.

Aspen back then was an unusual getaway in its halcyon natural setting. The designers who had assembled there shunned fads and shock value. They preferred the timelessness and simplicity of nature: the miracles manifest in apples and eggs and other compact packages that, seemingly minimal, encase great complexity. Bayer and Paepcke had created a place, away from the diversions and pressures of cosmopolitan living, where new ideals, true to the Bauhaus dream, might take root.

5.

Bayer kept the Bauhaus mentality alive not just at the Aspen Institute but in his own widely disseminated graphic design. Born in 1900 in Haag, Austria, he had gone to the Bauhaus in 1921, two years after the school was created, one year after Josef Albers arrived there, and one year before Anni started. All these individuals coming from a myriad of disciplines and locations were united by their wish to give humankind new functional designs. Bayer, like Josef Albers, was one of the gifted individuals who started as a student and became a master at the Dessau Bauhaus.

What Bayer was up to between 1928 and 1938 gets white-washed, however. He left the Bauhaus and moved to Berlin of his own volition in 1928. Most sources maintain the story that he eventually fled German repressiveness, which became intolerable to him after the Third Reich began in 1933. By the time he was the god of Aspen, deified as the perpetuator of Bauhaus design standards in postwar America, Bayer had managed to dispel rumors to the contrary. It is almost certain that the admiring Steve Jobs considered the guru of the new aesthetics to have been nothing but a victim of totalitarianism and a champion of liberty for all.

The years between the closing of the Bauhaus in 1933 and Bayer's flight from Germany five years later were never discussed.

In the United States, Bayer had further refined the streamlined vision he had developed with his new sans serif typeface and his geometric graphic layouts based on a grid but rendered playful and joyous. Bayer's lively graphics, devoid of historical reference and marvelously fresh, became part of the bread and butter of American life in the magazines *Fortune* and *Life* and in his logos

Herbert Bayer at the airport in 1969. He never talked about the period from 1933 to 1938, when he designed posters and books for the Third Reich to advertise and celebrate the glories of Nazi Germany.

and layouts for large American corporations whose products were everywhere in the country. His major clients included General Electric and the Container Corporation of America. Walter Paepcke was CEO of the latter—known simply as CCA—which is why Paepcke got Bayer to Aspen and selected him as architect of the building that housed the institute and also as its main graphic designer.

In 1938, Bayer had been invited to the United States by Alfred Barr, director of the Museum of Modern Art, to design and install a Bauhaus exhibition there. In helping select and then install this show at the Modern (in the days before it was ever called MoMA), he was among the first people to make the Bauhaus aesthetics integral to American life. His Aspen Institute further perpetuated the Bauhaus spirit. And that haven in Colorado gave Steve Jobs direct exposure to a legacy he already admired.

6.

Like Gropius, Bayer had immense personal allure. Josef Albers made a collage of two photographs he took of the young Bayer, shirtless and grinning ear to ear. His shock of straight blond hair, handsome features, and happy pose are those of a playboy with dash and mischief. Anni Albers, seeing this collage in the

Josef Albers made this collage of two photos he took of Herbert Bayer on a Bauhaus summer holiday in Ascona in 1930. The Bauhaus faculty members had a keen sense of fun and a playfulness that still surprises people. Bayer, with his dashing good looks, was every bit the charmer he appears to be.

late 1970s, shortly after Josef had died, smiled the way she did when sex and romance were on her mind. "Ahhhh! Zat Herbert!" she exclaimed. She went on to say what a good-looking ruffian he had been, irresistible to many ladies. Anni intimated that the real reason Walter Gropius left the Bauhaus in 1928 is that it was his only way of definitively halting the affair between Gropius's wife, Ise, and Bayer.

Bayer's capabilities as a lover were known by many. His politics were not. One has no idea how Steve Jobs, or Walter Paepcke, would have felt if they learned about Bayer's activities in Germany between the closing of the Bauhaus and what he always represented as his next step—flight from an oppressive regime. Bayer conveniently left out the five years of his life between 1933 and 1938.

Anni and Josef Albers stored two canvases by Bayer in a small utility closet in the void underneath the staircase of their house in Connecticut, where these pictures leaned against a hot-water heater. By the time I had met the Alberses in 1970, they had lost all contact with Bayer, who had given them the paintings in the 1950s. I assumed that the reason they did not store these paintings with the care they accorded to gifts from Klee, Kandinsky, and Schlemmer as well as from many of Josef's students, some entirely unknown, was because they did not like them as much. The main motif of Bayer's paintings was parallel lines in the form of curving brackets, like the pattern on certain Pepperidge Farm cookies. This soft lyricism was not the Alberses' sort of thing. But now I wonder if they had severed contact with their former colleague because they had become aware of Bayer's activities after the closure of the Bauhaus. They would initially have had no idea that Bayer played ball with the Third Reich. Following their own exile in 1933, they were absorbed in reformulating their existences at the fledgling Black Mountain College.

It is only recently that a lot of information has come out revealing the collaboration of former Bauhauslers with the Nazis. A pupil named Franz Ehrlich designed the entrance gates for Buchenwald. Another student, Fritz Ertl, was one of the architects of the gas chambers at Auschwitz. Brought to trial in Vienna in 1972, he was found innocent, because the gas chambers were identified as "shower rooms" on the blueprints of the camp, and there was no proving that they were anything else.

Shortly after arriving in the United States, Herbert Bayer attended *Bauhaus: 1919–1928,* the major Bauhaus exhibition at the Museum of Modern Art in New York in December 1938 and January 1939. That show got the American public to see the great achievements in every field of design in the school that had flourished before the then-current regime in Germany, soon to be America's official enemy, had shut it down brutally. Bayer himself had fallen out of favor with the German government, but only in 1937. He had a Jewish wife and did not support Nazi policy, but he had helped with the Third Reich's self-glorification with his posters and catalogues for the major exhibitions *Deutsches Volk, Deutsches Arbeit* (1934), *Das Wunder des Lebens* (1935), and *Deutschland* (1936).

Still, it is likely that Ertl knew that lethal gas, not water, would come out of the jets he positioned so meticulously. Most Bauhaus aficionados would rather ignore this. A Sunday *New York Times* article I wrote in 2009 about Bauhaus people who complied with the Third Reich incurred the wrath of many of the school's devotees who would prefer that these painful facts be forgotten. On the other hand, at least two art journalists have recently taken the Nazi connection in the other direction, exaggerating it by claiming that a former Bauhaus student was "the chief architect of Auschwitz."

Few people know that Herbert Bayer worked for the Nazis. The facts are clear. In 1936, a brochure Bayer designed for the new German government was distributed internationally. It was meant to lure people from all over the world to see the Führer's achievements. The Terramare Office—an official government-run public-relations organization whose task was to illuminate foreigners about the wonders of life in Nazi Germany—had commissioned this publication, entitled "German Youth in a Changing World." The cover brandished the image of a perfect Aryan: a blond-haired lad, smilingly earnest and proud, wearing the uniform of the Hitler Youth and holding a German flag. The photo was positioned jauntily with typography accentuating the pure, upbeat nature this fine young specimen of unadulterated Teutonic lineage embodied. It was just one of many such designs by Herbert Bayer, who created numerous brochures and posters for the Third Reich. On the single occasion when he was questioned about his work glorifying Hitler's Germany, Bayer commented only on the use of duotone technique and other formal elements of his work celebrating life under the Nazis. Otherwise, he successfully avoided the subject. His tactics worked; he was given a free pass.

7.

By 1983, when Steve Jobs gave his lecture there, the Aspen Institute's annual meetings had become the sine qua non of conferences for people at the forefront of design in every domain. If the attendees had known of Herbert Bayer's Nazi affiliations, they might have been less enthusiastic. But they were thrilled to be gathered at the mountain retreat and hear as fascinating and inventive an individual as the founder of Apple Computers.

To this audience of architects, industrial and graphic designers, teachers, and painters, Steve Jobs explained the functioning and purposes of personal computers. The listeners, while exceptionally bright, had little knowledge or understanding of these new devices. Jobs kept his description simple and direct. He emphasized the importance of the aesthetics of his products, and made clear his resounding passion to get their design right. But he acknowledged that it would be an uphill battle to achieve his dreams.

In his excitement over the objects he was developing for the benefit of anyone and everyone, as well as the frank apprehension that accompanied his enthusiasm, Jobs resembled Walter Gropius. Like the Bauhaus founder, he called on his audience to help make his venture successful. Promulgating his breakthrough ideas to people who had flocked to an institution created out of reverence for the Bauhaus, he spoke with a pioneering spirit and a willingness to hazard risks they cherished. The audience in Aspen was wowed.

For nearly thirty years, the precise text of Jobs's Aspen speech was presumed lost—or at least no one had succeeded in finding it. There is still no written transcript. But, in 2012, old-fashioned audio tapes turned up that provide a fuzzy recording of it. Listen-

Steve Jobs made a point of wearing a bow tie at the International Design Conference in Aspen in 1983 because of his $60 speaker's fee, but he still took off his jacket and rolled up his sleeves. After all, he was in action mode—pleading to the attendees to guide him toward making better-looking computers.

ing to Jobs in Aspen, you feel, as you could with Walter Gropius, the charisma that comes when a speaker cares with every fiber of his being about his mission. Jobs advocated computers that were sufficiently "adaptive," "simple," and "mundane" to be used by everyone. Like Gropius, he was not interested in products for the elite few, but in what could benefit the world at large.

Jobs explained that personal computers had first been developed in 1976. But he predicted that by 1986—within three years of when he was speaking—there would be ten million of them. His plea at Aspen was loud and clear: "I need your help. If you look at computers . . . they look like shit."

The attention that had been given to the appearance of automobiles, cameras, and watches—almost all of the best having been designed in Europe or Japan—had never been applied to

computers. Jobs saw this creation of better-looking computers as a task to be done in America. Like Gropius, he believed
that what would be an advance for all human beings could be
achieved only if many talented people worked collaboratively.
"We have a chance to make these things beautiful. . . . We really
really need your help."

In his straightforward speaking manner, the founder of
Apple, already well known as a successful inventor and businessman, gave further specifics on the history of computers,
how they functioned and how they could be improved, and their
graphic capacity in particular. He explained complex subjects in
clear, taut language that was easy to understand.

If there was anything pretentious about him, it was his deliberate unpretentiousness, his "I'm just an ordinary fella" affability.
Jobs had set the tone of his talk by opening with the statement
"They paid me sixty dollars, so I wore a tie." It was exceptional
for him to wear this symbol of respectability; he usually looked
ultracasual. The way he dressed most of the time suggested that
even if the objects one uses have to be meticulous, the fit of one's
jeans does not matter. Although hair did: either his shock of
thick black hair grew with extraordinary neatness, or he had
blow-dried and sprayed it.

Even though Jobs wore that tie for this exceptional occasion,
he had removed his sport coat and rolled up his sleeves. He was
determinedly affable. Like the hardcore faculty at the Bauhaus
(as opposed to Johannes Itten and a couple of other deliberate
bohemian types), Jobs wanted to be seen as a regular guy. To put
forward revolutionary ideas, you have a better chance of suc-
ceeding if the public does not perceive you as a weirdo. And you
have to express yourself so that your listeners feel respected, not
threatened.

Handsome, charismatic, and tireless, Walter Gropius led the Bauhaus to immense success, in spite of some fierce opposition and times when the school seemed on the verge of running out of money completely.

Jobs, like Gropius, used verbal language in a way consistent with the products he developed. His words are to the point; they elucidate ideas that others might obfuscate. Jobs was not as suave or worldly as Gropius, and lacked his notable physical handsomeness, but, like Gropius, he had the rare ability to bring his audience into the fold. Neither flaunted his exceptional intelligence in the manner of so many intellectuals and academics. So it would be with the material objects they were determined to create in multitude. Like the message and facts both Gropius and Jobs elucidated, the physical manifestations of their ideas were intended to put users at ease, to induce comfort rather than intimidate.

Jobs's smart and willing listeners in Aspen were well educated. Yet, after he asked how many of them understood personal computers, almost all raised their hands to declare that they did not. By their own admission, they were as unfamiliar with what he was about to tell them as were the people who went to hear Walter Gropius when he tried to spark interest and garner funding from audiences to whom simple functionalism was totally foreign.

Jobs, at least, had potential partisans. Gropius had the harder task of converting people who did not even want to consider design lacking a single rococo curlicue or Renaissance column. And Gropius invariably gave himself a second challenge. At the same time that he imparted the message that the Bauhaus would create objects that would give unprecedented ease and pleasure to the tasks of daily life, he was determining who was the most beautiful woman among his listeners and devising a means of meeting her the moment he stepped away from the podium. Jobs's personal passions were more confined and did not interfere with his thought processes when he spoke in public.

Still, even if the one had wandering eyes and the other thought only of his single mission, Walter Gropius and Steve Jobs had in common their capacity to coax their audience into territory they had never entered.

8.

Jobs's Aspen Institute talk did not present the computer as an amazing breakthrough. Rather, he fit it into a sequence of objects made to enhance people's capacity to communicate and to acquire information. He used a new vocabulary—with words like "motherboard" and "laptop"—that made the human body fundamental to these mechanical tools.

"Personal computers are a new medium," Jobs told that audience in Aspen. They were, however, a natural next step after the radio and, subsequently, the television. Television had reached a new level of effectiveness when the funeral of John F. Kennedy and then the launch of the *Apollo* spacecraft were viewed on screens all over the world. Those large TV screens were, Jobs pointed out, the ancestors of the smaller screens used for "Lisa-Draw," the oddly named application program largely forgotten today but that Jobs described to his visually sophisticated listeners as a breakthrough in computer optics. LisaDraw made it possible to draw on a computer screen and then transmit the image electronically.

Having sold twenty-five hundred Apple computers in 1977, shortly after developing the prototype in his parents' garage, Jobs had sold more than two hundred thousand Apple II's by 1980. But in 1981, the Apple III had proven to be the Edsel of Apple's product development. From Walter Isaacson's biography of Jobs, the go-to source for the inventor's basic history, we learn that, following the failure of Apple III, Jobs and his company hired a pair of savvy engineers from Hewlett-Packard to design a new microprocessor for them; it cost about two thousand dollars a unit. This was the "Lisa."

The Lisa also failed. It was too expensive. Three thousand that could never be sold ended up buried in a landfill in Utah. "Trial and error" is not unusual for revolutionary movements. At the Bauhaus, the greatest disaster was the chairs designed and produced, in quantity and at substantial expense, for an auditorium being developed in Berlin by the producer/director Erwin Piscator. They nearly caused the financial ruin of the Bauhaus. Piscator bailed out without paying a single Deutschmark for them. The mass production that was supposed to be the salvation of the Bauhaus's shaky finances ended up being another setback in Gropius's efforts to keep the school solvent.

"Edsels"—the product to which the Apple III was comparable—were a model of Ford automobile made between 1958 and 1960. Publicized as the car of the future, the "must-have" for all Americans, they were brazenly hideous. The Edsel's face had a vertical grille at the center, horizontal grilles to the right and left, and headlights like popping eyes. It resembled a bizarre sea monster. This car also had ridiculous sloping side panels and two-toned tailfins that made it look extremely proud of itself. And the Edsel, which advertisements flaunted as rust-proof, was so poorly made that it gave rise to a riff on its name: Every Day Something Else Leaks. If ever a mechanical object looked boastful, the engineered equivalent of a person who considers himself God's gift to the earth, the Edsel was it. It was the antithesis of everything Bauhaus; its appearance was inherently silly, conceived to embody braggadocio, and with no connection to function. (This vintage photo of it, below, makes the Edsel so

When the Ford Motor Company launched the Edsel in 1958, they expected it to be all the rage and to become a best seller. Instead, most of the American public loathed it, with its oversized features, and the Edsel was soon one of the greatest flops in the history of American business.

The Apple III, launched in 1980, was the Edsel of the computer company's product development. Clunky and awkward-looking, it failed to sell, causing Jobs to coax a pair of savvy engineers away from Hewlett-Packard to design a new microprocessor.

alluring today that some of us who scoffed at the car when we were children would love to have one now.)

Apple III and the Lisa were not as aesthetically problematic or mechanically flawed as Ford's car, but they embodied wastefulness. The ideal of all manmade products, from the Gothic cathedral to a desk lamp, was, according to the philosophical values that guided the Bauhaus, the effective use of minimal means. These computer models, failing to achieve that, were, like the Edsel, a setback for the company that made them, causing losses of both money and customer loyalty. (The Edsel lost the Ford Motor Company over 250 million dollars.)

But Jobs was proud of the new graphics developed for the Lisa. He wanted the people assembled at Herbert Bayer's brainchild

The Lisa, which Jobs named for the young daughter he barely acknowledged, was supposed to make up for the disaster of Apple III. But at two thousand dollars a unit, it also failed. Thousands of Lisas that were never sold ended up in a landfill in Utah.

to respect the achievement of the Lisa computer even though it had failed commercially. This financially unsuccessful model had made it possible to change the visuals of the computer screen to a far greater extent than ever before. The capacity to shade lines and vary the background was a breakthrough. Previously, only a single font appeared on a computer screen; now there were choices. Enumerating these advances, Jobs was striking a note that he knew was of central importance to the Aspen audience.

Isaacson points out that Lisa was the name of Jobs's sole child at the time. Jobs eventually told Isaacson that there was no question that he had named the computer for the daughter he had initially failed to acknowledge. Similarly, the Edsel was for Edsel Ford, son of Henry Ford. Perhaps this is a cautionary tale about entrepreneurs naming products for their children; when the product fails, the gesture assumes an unhappy meaning.

In 2018, Lisa Brennan-Jobs published a book that tells the story of her parents. They were high-school sweethearts who lived in a cabin together one summer, and two years later in a modest house in Cupertino, California, where Lisa's future mother, Chrisann Brennan, worked in the packing department at Apple.

Brennan had left Jobs before she realized she was pregnant. Once the baby girl was born, in May 1978, he denied that she was his. Lisa's mother worked as a house cleaner and waitress to support herself and the baby. She also depended on welfare payments. Then the district attorney of San Mateo County sued Jobs on her behalf for child support. He denied paternity, claiming he was sterile, but in 1980 a DNA test proved otherwise. Jobs, by then worth hundreds of millions of dollars, assumed modest financial responsibility for his daughter.

9.

After charting Apple's design failures, Steve Jobs lapsed into West Coast–speak. "During the next fifteen years, we have an opportunity to do it great or do it so-so."

Presumably Jobs knew that the use of "great" and "so-so" where adverbs were required violated the rules of correct English grammar. Within a few years, Apple Computer would catapult itself out of yet another case of the doldrums with the slogan "Think different." This new, staccato style of language said, in effect, "Screw tradition! Break the rules!" The directness was the same form of rebellion that fueled a lot of the Bauhaus pedagogy and design. So was the idea of teamwork. Jobs told the packed hall in Aspen that, for three and a half years, everyone who worked on Lisa had joined forces to improve its programs.

The collaborative effort to achieve a mutual goal was identical to the approach of the students and teachers at the Bauhaus. Previously, it had been impossible to embed graphics. Moreover, the "*i*'s were as wide as the '*w*'s." The Lisa had succeeded in "injecting some liberal arts into computers." The new capabilities enlarged the population of computer users from a select few to anyone writing a paper at high school or college.

Jobs told the Aspen audience that the first computers had been large and awkward, and only a few people knew how to use them. Then, in the 1960s, time sharing started, with groups of people co-owning a single computer. When personal computers were developed in 1976, they transformed human communication.

Jobs emphasized his perpetual goal of sticking to basics. Computers needed to be easier to use. Their mechanisms belonged to an era of rapid change and progress, but they depended on universal laws. Video games, for example, followed the laws of gravity and momentum. They simply adapted known phenomena and long-recognized truths to new technology. Jobs emphasized that the wisdom of the ancient Greeks remained totally pertinent today. It was the identical message taught at the Bauhaus.

Jobs said that at school he had learned about Plato and Aristotle through the medium of books. Yet, for all the benefits of his reading, he regretted that "there was nothing in the middle." He hypothesized: if computers were developed perfectly, you would be able to ask the question "What would Aristotle think?"

Invoking Aristotle, Jobs was implicitly championing the notion of inquiry. This was at the heart of computer usage. He likened Aristotle's mind to that of a computer in its capacity to address a wide variety of subjects, including politics, economics, psychology, music, theater, zoology, biology, physics, logic, and ethics. Jobs said that human beings would use these new devices in accord with Aristotle's emphasis on perception and empiricism. Notions of absolute truth did not pertain.

Then Jobs spoke about Aristotle's teacher, Plato. Plato was the Greek philosopher most often cited at the Bauhaus. With Plato's name, Jobs was conjuring the person who considered well-conceived forms to be the blueprints of perfection. Plato emphasized appearances, the importance of optical vision, and the miracle of sight. Looks mattered immensely.

10.

The Aspen crowd robustly applauded Jobs's twenty-minute-long talk. People then asked questions. Unfortunately they did not use microphones, so what they asked was not recorded, but Jobs's answers to them were. He explained that Apple had a program to donate a computer to every school in California—a total of sixty thousand. They were working to "make computers easier to use," and since a hundred percent of Apple's employees owned stock in the company, with no differentiation between labor and management, everyone had the "same goal and objectives." This further emphasized the spirit that made Gropius liken the Bauhaus to a Gothic cathedral. Community effort, with a purpose shared by all the participants, was essential. Jobs described the way the Apple team all pursued their objectives in unison, avoiding internecine competitiveness and conflicts. As at the Bauhaus, communal meals were part of the routine; people who dine together are more likely to get along well at work.

"What we can give them in the next five years is a lot." If the mechanics were adequate, the computers of the future would give people a chance to imagine asking questions of Aristotle or

Plato and having answers approximating responses the Greeks might have formulated. But these new computers had to look right as well as function correctly. "We have a shot at putting a great object out there; if we don't, we are putting an ugly object there."

In Aspen, Jobs was addressing an audience he respected more than dry engineers or technicians. "Computer people are closest to artists than anyone else," he said in a voice both humble and pleading. Like Walter Gropius, who summoned Paul Klee and Wassily Kandinsky to his new design school, Jobs believed that it would be a handful of artistic geniuses who would enable him to create tools that would truly ameliorate the human condition. They needed to do so in a setting akin to the Bauhaus: "an environment where they don't have to convince others, where all believe that they are making a difference." Jobs had the same optimism as Gropius in believing he could create that utopia.

11.

Several sources report that in his 1983 talk at the International Design Conference in Aspen, Steve Jobs spoke about the Bauhaus and his adulation of it. The initial basis of this idea is Walter Isaacson's book, which was published in 2011. Isaacson describes Jobs's reverence for Bauhaus values in a way that has made Apple aficionados infer, incorrectly, that Jobs referred to the Bauhaus by name, although Isaacson does not specify that. In any event, it was only in 2012 that the recording of Jobs's "missing lecture" in Aspen was found. Jobs never once refers to the Bauhaus specifically. Nor does Jobs ever name a Bauhaus artist—even Herbert Bayer, who had created the haven that was his pulpit. But, as he often did, he talks about ideals that can be related to the Bauhaus. And there were many other occasions when he praised the Bauhaus by name for having simplified and humanized the material objects of everyday living.

Steve Jobs, in fact, was responsible for the core objectives of the Bauhaus ultimately being realized. What was launched in Weimar in 1919 would only fully blossom half a century after the Bauhaus closed. It would only be when the stalwarts of the great design school were no longer alive that the object that embodies their dream would be born in a garage in California. From there, it would multiply to every corner of the world, as they intended.

This tool that encapsulates Bauhaus theory has become a

Sleek and streamlined, the iPhone uses modern materials and machine technology to great effect. Its design is unique to its era, without a hint of older styles or fashions. Everything that one sees is there for a purpose—the inevitable Apple logo being the only exception. Nearly a century after the US Supreme Court debated whether a Brancusi sculpture was a work of art or a propeller blade, and therefore whether its importation was taxable, there is little distinction between the appearance of a useful device and that of an abstract sculpture.

staple of human existence, constantly serving everyday needs while affording its hundreds of millions of users the satisfaction of its honest and pleasing aesthetics. A lot of the advances in human communication and information-gathering vital to the iPhone pertain to a vast range of modern devices, but the iPhone puts them in their ideal package. Many of the attributes, aesthetic and functional, explored in this book, exist in a myriad of smartphones in general and in other incarnations, but the basic iPhone is, visually, the perfect Bauhaus design. Its means of manipulation also honors Bauhaus standards.

That fealty is no surprise. From the moment he had learned about the experimental institution with its resident geniuses who had come from all over to be at the epicenter of new forms of creation, Steve Jobs had immersed himself in the rigorous beliefs of the Bauhauslers, and studied their approach to physical form and to human need. He then fulfilled the goals of the Bauhaus with the iPhone.

Part II

1.

No one working for Apple Computers was aware that I was writing this book until I signed off on the manuscript. I might have learned significant details if I had met individuals involved in iPhone design or fabrication, but I did not want to be either influenced or disappointed.

With my premise that the iPhone is the ultimate embodiment of the Bauhaus goal of good design applied to mass production, I avoided any chance of being swayed by people with vested interests.

An important caveat: it is the Bauhaus as a propagator of design standards and goals concerning the appearance and functioning of everyday objects that lives on in the iPhone, not the role of the school as a paradise for artistic geniuses. Several individuals of unparalleled imagination and poetry were succored there. Nothing about the iPhone rivals their achievement. Gropius's purpose of the development and dissemination of handsome, useful tools for human betterment is the aspect of the Bauhaus where the iPhone fulfills the ideal.

If you are looking for an analysis of all that the iPhone has achieved from a business point of view, this book will not provide it. Nor does it address the chicanery behind the elements of its functioning that force the consumer to spend more money. This text will not enlighten you about the complex politics inherent in the technological revolution. It does not make feints into a lot of territory central to the story of the iPhone: the mechanical functioning of the device or the nuances of its relationship to BlackBerries and the vast range of other modern communications and smartphones. While it considers the unique viewpoints of some of the individuals who steered Apple Computers in the company's own distinct way, it does not pretend to offer the slightest expertise in economic or anthropological analysis, or try to be another of the many histories of either Apple or the companies like Google with which its products are interdependent. Its purpose, rather, is simply to make clear how the iPhone incarnates certain goals and values of the pioneering art school that flourished briefly in Germany in an interlude of exceptional human progress between two world wars as a creative haven that, in spite of the evils smoldering in the society around it, changed the visible world forever through its standards for mass production.

The iPhone adheres to all the principles of Bauhaus design at its ultimate. This flat tool of communications, compact and clean and easy to hold, reflective of the teamwork of professionals in many fields, relatively inexpensive and accessible to one and all, has, for better or worse, made the Bauhaus's reigning principles as universal as its guiding geniuses hoped they would become. The smooth corners of its rectangular form render it agreeable to use; its plastic, glass, and aluminum casing, all calmly monochromatic, are pleasing to the touch. It fits comfortably in the palm of the human hand, and responds rapidly to contact with one's fingers; it is what we call "user-friendly."

The iPhone's thinness and scale make it easy for people to hold; the keyboard responds to touch and requires no expending of physical energy. This emphasis on usefulness to human beings, in both the physical object and its many applications, was a core objective at the Bauhaus.

It is conveniently lightweight, and small enough to fit into most pockets.

Of course it imposes nightmarish experiences. Apple Stores are best avoided. The packaging on Apple products—self-important, difficult to open, and, in spite of a bow toward political correctness, often ecologically wasteful—is irritating.

Still, the iPhone itself functions in harmony with the world around it, being visually unobtrusive and occupying little space. It is, at the same time, large enough to be legible. Its predominant colors are an appealing shiny black, a pure and unadulterated white, and the matte gray of its anodized aluminum back. It does not have an ounce of gratuitous ornament. We accept the representation of an apple on its back—unobtrusive, neither raised nor recessed, demarcated only by the variance of texture—as integral to its being, a form of identification and a source of emotional lift, not decoration. It is legible, totally, but nothing about it announces itself. Its construction is impeccable, and usually its functioning is as well. It reflects immense intelligence and forethought, and one feels the invisible labor that has made it what it is.

Not that this product made by the Bauhaus aficionado Steven Paul Jobs is flawless, even as its looks and capacities and widespread use have taken Bauhaus theory and vision to the stratosphere. The iPhone causes no end of frustration to many people when it malfunctions or runs out of charge too quickly. Moreover, its detrimental effects on everyday human existence often outweigh its advantages. Little in modern life is more pathetic than the sight of families dining together with everyone looking only at his or her mobile phone, be it an iPhone or another model. They engage only with it—or, *through* it, with someone who is absent—rather than being present with the other people who are actually there. Individuals strolling in glorious countryside or on the pathways of Venice sending or receiving text messages, stooped over the instruments in the palms of their hands rather than looking at the world around them, are grim and distressing. To wheel one's baby in a stroller while using a phone rather than experiencing the wonders of life itself is a loss for both parent and child.

But, then again, an object cannot dictate its own use. If someone feels so relaxed and confident in a Wassily armchair—one of the greatest of all Bauhaus designs, with its support structure like a tensile drawing in space, its back and seat an intelligent and elegant use of leather or canvas or horsehair pulled taut—that he or she starts to shout violently, it is not the fault of Marcel Breuer. He designed it to enhance, not thwart, the pleasures

Smartphone use has had an indelible impact on human behavior. Not only are this young woman and man, embracing each other physically, distracted from what might be a splendid moment to be savored totally, but they are not even looking at one another. This sweet, ironic photo depicts the tragedy of people possessed by their possessions.

The iPhone has changed the nature of communal encounters. It gives people the option to eat and drink together in such a way that, while they are physically united, they engage as they choose in their private territories. Whether they are communicating with other people far away, playing games that entertain them, or reading about subjects that interest them more than speaking to one another, it fulfills a wish to be simultaneously in the presence of friends but in a different universe. Some observers consider this an advance in human camaraderie; others deem it a setback to the nature of personal relationships.

Marcel Breuer's armchair, with its horsehair fabric stretched taut on an armature of tubular chrome, is one of the Bauhaus designs that has remained a classic. Like Josef Albers, Breuer arrived in Weimar as a student and then became a teacher. He made this chair in 1927 when he was head of the school's furniture workshop in Dessau. He got the idea of using hollow tubing from bicycle handlebars.

of reading or of human congress. If one of Marianne Brandt's refined stainless-steel table lamps, with its luminous glass globe and perfectly proportioned vertical support and round base, facilitates the writing of hate mail, neither the designer nor the product is to blame. There are people who are owned by their iPhones—who have become their victims rather than their pilots—but that is not the fault of the instrument or its crafting.

Still, this is not a book about iPhone uses, which are nearly infinite. Its capacity as a pocket-sized camera has changed how people capture and transmit information by enabling them instantly to have visual records of moments they want to remember, of objects they are considering buying, or of key information and facts more quickly photographed than written down. Smartphones of every sort facilitate instant communication so that someone standing at the vegetable stand can ask people fifteen miles away if they want peas or string beans with dinner; these small devices also combat loneliness by providing constant company and making individuals feel perpetually in touch with one another. These immense capacities, as well as the potentially negative consequences, have already inspired, and will continue to prompt, copious commentary. The premise of this book is solely that the iPhone has become essential to hundreds of millions of people in the conduct of their professional, social, romantic, and family lives; the standards and aims of the Bauhaus, not totally realized when the school was running, live today in every corner of the globe. Not that iPhones are modern design masterpieces like the Jaguar XK120, the Chemex coffeemaker, the Olivetti Lettera 22 typewriter, or the creations of Dieter Rams and Raymond Loewy. But they have sufficient qualities to be the epitome of Bauhaus functionalism and grace.

2.

I see the iPhone as the ultimate realization of the Bauhaus's dream for the world with the good fortune of a rare perspective. In the 1970s, I devoured Bauhaus values at the source. I bridge the gap between two great epochs in the modernization of civilization, since, in the period following the deaths of Anni and Josef Albers and everyone else who was actually at the Bauhaus, I have witnessed the revolutionary development through which personal computers have transformed everyday life. Now that those devices have become as small as playing cards—in the forms of iPhones and their clones used by billions of people—I believe that Anni and Josef would agree that the iPhone is the manifestation of their deepest beliefs concerning the stuff of human existence.

Josef Albers was at the Bauhaus longer than any other single individual, having arrived in 1920 and remained to the bitter end; Anni was part of the great institution for almost as long. They were close to Gropius and his first and second wives, to the Klees, the Kandinskys, the Schlemmers, the Breuers, Mies van der Rohe and both his wife and his mistress, Bayer, and the other principal personages of the institution.

The Alberses were not nostalgic; what interested them was their current work, which was perpetually evolving even as Josef was in his early eighties and Anni in her early seventies. They refused to glorify "the good old days" in Weimar and Dessau, which they knew had been rife with conflict and struggle. But they adhered rigorously to the philosophy that Gropius put at the core of the school. They maintained its abiding faith in the power of the tangible to enter the soul. They revered good form, in behavior and objects alike, in accord with the ideas of Plato, the Bauhaus's intellectual god.

In their own manner, the Alberses were still living the Bauhaus ideals. As our relationships evolved, Anni in one way and Josef in another let down the barriers and talked to me about the essence of life and work in both Weimar and Dessau. They spoke of Gropius's dream and what it meant in the most universal sense.

They also gave glimpses, in novelistic detail, of the personalities of fantastic people. Nina Kandinsky was the coquettish wife obsessed with her own youth, while her older, brilliant husband was obtusely intellectual. She batted her eyelashes and flirted while refusing to tell her age to a policeman who stopped her for riding her bicycle against the traffic, Anni recalled with her mischievous grin. Wassily, meanwhile, developed theories of the sounds emitted by different colors, a subject he explored with Josef. The Alberses made Paul and Lily Klee and their son, Felix, and their cat, Fritz, come alive as a family both eccentric and astonishingly like everyone else in their domestic needs. Paul Klee was as capable at arguing with a train conductor about the reason Fritz did not need a special feline ticket—in the third-class carriage they took on a family holiday to the beach—as he was at painting masterpieces. Lily was often away being treated for neurasthenia, while it was Paul's task to get dinner on the

Anni and Josef Albers maintained the truest Bauhaus values lifelong. They loathed nostalgia and always lived in the present, but their faith in art and design as forms of service and celebration was true to the ethos with which the great design school had been founded. Henri Cartier-Bresson, an admirer of both the Alberses, took this photo of them at home in Connecticut in 1968. Their mix of gentleness and strength, the force of their quietude, and the rapport between them are salient.

table at a reasonable hour every night and oversee the teenage Felix's progress as the youngest student at the Bauhaus. Alma Mahler Gropius was like a sorceress, and Lilly Reich a caustic and duplicitous tyrant. What became clear was that the stars of the Bauhaus were, while brilliantly creative and intrepid, as human as everyone else. And all of them were obsessed with the appearances of the objects that served their quotidian needs.

Once the Alberses had the confidence that I was not just a hanger-on eager for snippets from their colorful past, that what excited me was the centrality they accorded art and design in all of human existence, they became relaxed enough to share their rich memories of life at the Bauhaus. They demonstrated and discussed how the personal and the aesthetic and the material are interrelated rather than separate. Josef and Anni illuminated the way that morality is intrinsic to the visual.

Anni and Josef were insistent that the essence of the school's belief system and practices was not to be found in expensive imitations of Mies's glass-topped coffee tables and in readaptions of Brandt's gorgeous but costly metal serving bowls. They had no use for what Bloomingdale's and other high-class merchants were now labeling "Bauhaus style." For them, what counted was that the visible embody an attitude of honesty and integrity and guilelessness.

The Alberses, like their best-known Bauhaus cohorts, cherished no-nonsense yet felicitous design as a vehicle of decency and a deliberate rebuttal to artifice. It must never be about the self, or express what should be kept personal and private. It had to be geared to the universal.

3.

The Bauhaus fostered a faith so powerful that it lured like-minded souls from entirely different backgrounds. Josef Albers was a schoolteacher in an agricultural village where the locals called themselves "peasants" and then in a small mining town. Anni was from cosmopolitan Berlin where people packed as many steamer trunks for their holidays in Baden-Baden or the Côte d'Azur as they needed to hold their wardrobes for daytime sport and the evening dress requisite in elegant dining rooms; there was always staff to carry the luggage. Mies van der Rohe grew up in the school of hard knocks in Aachen. A bricklayer's son, he was used to the police breaking up his fights with other tough local lads. Gropius came from landed gentry, aristocrats as well as successful professionals, the men all highly ranked in the military. Several of the women in the weaving workshops were from Eastern European villages where their mothers did handiwork to supplement the income of their laborer fathers.

After the Third Reich forced the Bauhaus to close, the Bauhauslers dispersed to everywhere from America to Israel, with the Klees going to Switzerland and the Kandinskys to Paris. They continued to live the values the lot of them shared, however different their other choices in life. But none were as steadfast in maintaining Bauhaus simplicity and the school's lean aesthetics as the Alberses. Their Connecticut home, which is where I met them, was a true Bauhaus workplace governed by the Bauhaus standards of which the iPhone would ultimately be the quintessence.

Josef, even as an octogenarian, was tireless in his daily creation of the *Homage to the Square* series as vehicles to present color relationships and effects. He was also perpetually devel-

Annelise Elsa Frieda Fleischmann—the future Anni Albers—grew up in a luxurious milieu where the expectation was that, like her mother, she would marry, have children, run an affluent household, and do nothing else. Her parents' dining room, like the rest of their flat in a fashionable Berlin neighborhood, had nothing modern in it except the concealed electric buzzer to summon the help. When Annelise told her father about the Bauhaus and the functional modernism it was promulgating, Herr Fleischmann replied, "What do you mean, a 'new style'? We have had the Renaissance. We have had the baroque. There is nothing left to do."

oping configurations of straight lines that deceive the viewer in their apparent readability as plastic forms, with walls and openings, only to become something entirely different as we look at them. Physical impossibility in the guise of the plausible was his elixir. For Josef, the goal of art was to reveal new ways of seeing, "to open eyes."

He became internationally famous and financially successful, but that was not what counted. A revered and influential art teacher, an artist whose paintings sold at hefty prices and whose sculptural reliefs were commissioned by top architects and seen by vast numbers of people, and the subject of television documentaries, he enjoyed these hallmarks of recognition, but what mattered was his mission, not his celebrity.

Josef focused on inventiveness and out-of-the-box thinking that dealt with universal issues and needs in new ways. His quarry was mostly in the realm of line and color. Steve Jobs had a similar agenda—in the arena of communication and access to information. When Josef Albers said that he, like the greatest of his cohorts at the Bauhaus, had greater respect for good household appliances or well-designed typewriters than for paintings "like those of Cy Twombly or Jasper Johns," it was not out of animus but, rather, out of heartfelt belief. Like Steve Jobs, he prized art and design that were accessible to everyone, and had no patience for obscurity or elitism. Art that spoke only to sophisticates was his nemesis. What communicated to one and all, and brightened the days of people from every walk of society, was his abiding passion and his personal quarry.

Josef had a particular fondness for photocopy machines. He studied their appearance and workings almost every single day when he went to the local "copy shop" to make multiples of letters he had written or of the latest magazine articles about him and Anni. He did not seem to notice that Mrs. Fago, the woman who ran the shop, was as snoopy as a spy, and far less subtle. She kept track of his most private and sometimes very personal communications as if it was her business not just to know things but then to report them to me when I next went to her workplace at Josef's request. In fact, his imperviousness to that sort of human boundary breaking or invasiveness went hand in hand with his respect for those IBM machines that could not only provide instant replicas of printed matter, but even do so in color.

Tools are, at their best, reliable. With simple ones, like hammers and pliers, you see all the working parts. With more complex ones, like photocopiers, the inner mechanisms are concealed from view, but all the components in tandem achieve a single goal.

Coordinated movement of different parts, all in a sturdy casing, gave Josef a feeling of trust. They reminded him of bees building their hives and producing honey on meticulous combs, of entire communities in ancient Mexico and Egypt constructing pyramids where later generations completed what earlier ones had begun. Their purpose was service. By contrast, abstract expressionism and work by artists like Francis Bacon were too personal and focused on the self. To make matters worse, they were messy. Well-designed objects, on the other hand, embodied and transmitted rightness.

4.

The Alberses professed and lived with the same ideals Walter Gropius had championed to the artists and artisans of the Bauhaus. They were convinced that the human soul benefits directly from the objects we see and use, by conveying feelings of rightness and ease and optimism—as opposed to irritation and falseness.

Anni's professional territory was in weaving textiles. She celebrated what was inherent to cloth and every conceivable composition, natural or artificial; the means by which they could be crossed over and under and through one another to construct pliable material were her elixir. Functionality was essential. When her task to earn her Bauhaus diploma was to devise a wall covering for an auditorium, its visible side reflecting light and its concealed one absorbing sound, she embraced the challenge. Not only was she awarded the diploma that had surprising significance for someone who shunned most measures of achievement, but over three hundred linear meters of her material was used with great success in the Federal Trade Union Headquarters, a dashing modern building near Berlin designed by Hannes Meyer, the second Bauhaus Director.

Anni admired synthetics. She uttered the word "Polyester" as if chanting "Hallelujah." Success at their tasks made her preferred materials reassuring, even noble. Gratuitous decoration or self-proclaiming artiness—"you know, those dresses that look like nomads, tents in an Arabian desert"—manifest falseness, affect rather than authenticity. Her taste prevailed in what she made for the world and in all that she chose for herself.

One of the primary tenets of Aristotle's approach to how we come to know or feel things is that a lot starts with what we "sense"—which is to say how we perceive things. We begin with an instinctive response more than through an acquaintance with the facts. Gut reaction often precedes knowledge.

I want you to feel the aspirations and achievements of the Bauhaus as if you, too, had the chance to enter the pristine world of Anni and Josef Albers. You will discover that simplicity can be as rich as it is austere. The whiteness, the intelligence, the refinement, the apparent modesty, the radiating genius, and the sense of total originality envelop you.

If you are lucky, you have known similar feelings just walking into other fantastic places. The sheer awe of entering the Chartres cathedral penetrates your being. The glory of King's College Chapel in Cambridge, England, fills you with hope. The humor and frippery and guts and imagination of Le Corbusier's church at Ronchamp makes you want to do a jig. The sight of dawn breaking at Machu Picchu gives you a sense of the eternal. There are many such moments when architecture enters your heart and soul, and you feel a consuming uplift; such occasions are among the marvels of human life, and result from the hallmarks of what our fellow people, as opposed to any other sort of animal, have achieved.

Of course other forms of dwelling are equally miraculous, even if their direct emotional effect on us is less dramatic. Beehives, termite hills, birds' nests: these carefully conceived uses of material, delicately assembled, brilliantly functional, never fail to amaze. This is true of spiders' webs, those incredible weavings of hair-thin naturally excreted filament that are put together in a graceful openwork not to be a home but in order to trap insects as food.

But the impact of entering the house of the two brilliant Bauhauslers who outlived most of the others, and who maintained the spirit of the institution in the nuts and bolts of their marriage as in their art and their writing and the values with which they lived, is particular. The heavenly simplicity inside 808 Birchwood Drive was made all the more striking because of the humility of its presentation. In certain ways, the Alberses' house had the quality of old-fashioned hardware stores. It was as spare as could be, while those emporia are packed to the gills, but it provided the same feeling of honesty. A change in breathing and inner rhythm, and a complete sense of well-being, occur in those settings where candor and dependability reign. Authentic hardware stores have well-sanded wooden floors and an ambient aroma that combines their subtle smell with the scents of furniture wax and grass seed. The people who help you in them are instantly trustworthy. They are willing to give advice on how to join a corner or buff a steel post, and explain this succinctly and accurately. These marketplaces of screws and

paints and garden equipment and other goods to aid the mainte-
nance of everyday life, with their power tools and synthetic res-
ins and electric kitchen tools, have a feeling completely different
from the large building-supply warehouses of today, with their
predominant smell of plastic. There is an atmosphere of grit.
The pervading sense of the way things work provides feelings of
rightness and of sheer reality. It is the "this-leads-to-that" you
enjoy when you bite an apple or hit a tennis ball on the sweet
spot of the strings. Such hardware stores are oases of wholeness
lacking in many modern institutions, and the Alberses' house
had that same sense of the joy of work and a total absence of
trickiness.

The household of these two last Bauhaus gods was Spartan.
There was nothing whatsoever on the coffee table composed of
a thin gray marble slab supported by a plain white metal frame.
Josef's desktop, a simple wooden plank, had nothing on it.
The only objects on the otherwise bare kitchen counters were a
Salton yogurt maker and a chrome-plated toaster. A small portable television sat on a rolling cart next to Anni's bed. The Polaroid SX-70 camera lived on the Formica-faced office cabinets
that served as storage for clothes in Anni's bedroom. These were
sacred objects: the equivalent of bronze gods and goddesses or
wooden Jesuses in other houses. The Alberses had a fantastic
collection of Mayan and Aztecan and Incan corn goddesses and
representations of Tlaloc, the Mexican rain god, stored in a
closet next to their dryer and washing machine, but hardly anyone else knew they were there. In other households, this bare-
bones storage behind cheap, hollow-core wood doors would
have been used for cleaning supplies. The few appliances that
were visible in the rest of the house were the modern equivalents
of these concealed ancient deities.

Not that Anni and Josef understood the technology of the

objects they admired. The Sony television, less than a cubic foot in overall size, was among the devices they most often cited as "a real work of art, so much better than those people trying to express themselves by hanging their personal laundry out in public, disordered and messy." When there was a CBS special on Josef and his art, they sat together on Anni's bed so that they could both watch. Anni liked to watch television—especially *The Merv Griffin Show* (she pronounced his name "Me-rhev Gliffin," with palpable delight, since she loved his manner and style) and the Watergate hearings and tennis tournaments. (This was the era when players still followed the "all-whites" rule.) Josef hardly ever watched TV, which is why it was in Anni's bedroom. But the CBS special was an event to share.

So on Sunday morning, January 2, 1972, they tuned in to *Camera Three* on Channel 3, pleased to know which local station was affiliate with which of the three major national networks. They liked the way that each network was identified by only three letters, each with *B* in the middle. Josef had been told by the art dealer Sidney Janis that Frank Stanton, who ran CBS, was a fan of his work and owned two examples of his *Homage to the Square*. The stars seemed perfectly aligned as they waited for the moment when the show would begin. Anticipating it, they were fascinated by the advertising for aspirin and deodorant as reflections of general human needs.

The half hour devoted to Josef and his art began. The critic Grace Glueck interviewed the painter warmly. Glueck, who wrote for *The New York Times,* shared Josef's love of potato pancakes. She had entitled a piece about his *Homages* "Each Day, Another Albers Pancake." She knew her subject well, and on the multiple occasions she had visited the house with a film crew, she had elicited his sense of fun as well as his rigor.

Now the Alberses were watching those encounters in the very place into which viewers all over America were simultaneously being brought by the tool of television: the device Steve Jobs considered to be the first great exemplar of making the world accessible to billions of people simultaneously.

There was a problem. The Alberses had been told from the start that the show would be in color. This was vital for Josef's work, and it was one of the reasons he had agreed to do the program. But when the Alberses tuned in, there was no color. They moved the little antenna on top of the TV, trying to invoke the yellows and reds they had been promised by the film crew. Still, everything remained black, white, and gray.

After five minutes, Anni telephoned Laurel Vlock, a local TV host who had commissioned a rug from her and was a helpful and sympathetic friend. Surely Laurel would know what to do.

"Oh, Anni, I suspect that the problem is that you only have a black-and-white television. You need a different model." The Alberses had no idea about any of this.

Steve Jobs had a different issue with Sony products. Initially, as a young man, he had admired them greatly. Then he became bothered by their pervasive blackness. His decision to make products that, by contrast, were white—less like automobile tires and more like the building that housed the Bauhaus— would change modern design.

5.

To Josef Albers, rigor and visual neatness were imperative to great art, as was its service to a purpose. Even if Frank Stella's lines were clean enough, they moved nowhere and had no effect; if Robert Rauschenberg's work had a certain verve, it was, for

Albers, disorganized—although Rauschenberg had been his student and often told others of the indelible effect of Albers's teaching. Grace and smooth functioning were imperative. "Max Beckmann—a schwindel!" Josef would say, always pronouncing "swindle" with an extra whoosh. "He doesn't know how to put colors next to one another; always those black lines!"

Josef went into these diatribes regularly. He was adamant; these were life-and-death matters. They pertained to the need for art to be generous to all who looked at it, to celebrate the wonder of life itself and not to be connected to the ego. But as often as he blasted artists who he felt tried too hard to appear bohemian or devoted too much energy to playing the right cards for success in the art world, he invariably followed the declamations with the highest praise for what he loved. "You see that Polaroid camera," Josef would say. "I prefer great machines to poor art. It's handsome. It brings something into the lives of all its users." The Alberses had bought an SX-70 shortly after that breakthrough model came onto the market. "Now, that is so much better than lazy art by self-infatuated people. It makes miracles happen. You take the picture, and a minute later you have it printed."

Albers had been innovative with his Leica when he was at the Bauhaus and, later on, traveling in Mexico. He had also relished the visits of Henri Cartier-Bresson, Lord Snowdon, Yousuf Karsh, Arnold Newman, and other gifted photographers when they came to Connecticut to make portraits of him, and liked discussing their cameras. The mechanics of the Polaroid with its instant on-the-spot developing process were an entry into paradise. Josef and Anni's Mercedes-Benz—the only major luxury in which they indulged—was yet another example: a well-functioning machine, designed intelligently, affording pleasure to the user. Praise for these sublime machines from Josef would

The Polaroid SX-70, designed and produced by the brilliant Edwin Land in 1972, could develop pictures on glossy paper in sixty seconds. The Alberses bought one almost as soon as these cameras became available. They considered it an artistic masterpiece as well as a thrilling addendum to their everyday life. They admired the materials and the functioning of the camera, from its stainless-steel-and-leather shell that folded flat and opened for use to the glossy paper that would emerge from it with the photos freshly printed.

often precede another storm, this time the butt of his anger possibly Robert Motherwell or Morris Louis or Jules Olitski.

In those beliefs, Anni and Josef were like a two-person religious sect. The Bauhaus had been the place where they had met and started to practice their faith. They were still living according to its standards when I met them in 1971. The respect for the marriage of form and function, the love of simplicity and efficiency, the faith that good design improved one's inner life, constituted the rallying cry about which they had first bonded when they met in 1922, and it was every bit as dear to them half a century later.

The twenty-two-year-old Annelise Elsa Frieda Fleischmann,

Tupperware, invented in 1946 by the American Earl Silas Tupper, became one of Anni Albers's favorite things. She said it embodied "the true spirit of the Bauhaus." She admired the technology behind it, and the lack of gratuitous ornament. It was designed for maximum effectiveness as well as visual allure. Anni knew nothing of Tupperware parties, but admired this inexpensive kitchenware and most anything else that was functional and free of decoration, especially a container that could be sealed airtight, when she went what she called "treasure hunting" in her local Sears, Roebuck.

from a wealthy Berlin family that was Jewish on both sides yet had converted to Protestantism, and the thirty-four-year-old Josef Albers, a devout Catholic from a poor family of laborers in the industrial and coal-mining city of Bottrop, in the Ruhr valley, were united by what mattered most to them. Intelligent reductionism and modest presentation were essential. Pretentiousness, whether intellectual or social, was deplorable. The cement of their fifty-year-long marriage was evident when they delighted in the capacity of their Chemex coffeemaker—made by Josef's Bauhaus friend Peter Schlumbohm—or their electric yogurt maker, which was so easy to operate and clean. This was before the era when grocery stores stocked yogurt of every variety. They exulted in the way the simple machine, mainly white plastic, transformed milk, and delighted in the qualities of the results. Whiteness and purity were unimpeachable sources of joy.

6.

In 1972, I took Anni Albers to my family's printing company, Fox Press, which was just north of Hartford, Connecticut.

On the first occasion of my meeting the Alberses, the previous year, Josef had asked me, "What do you do, boy?" I told him I was studying art history at Yale Graduate School. He was pleased when I answered his question about whether I liked it by explaining that I really did not. I told the former Yale professor staring at me intensely that I felt I was losing my passion for looking at art while being made to pursue unimportant information. I had become outraged when my professor in a course called "Seurat and the Iconography of Entertainment" had answered my question about how Seurat put points of color

When Annelise arrived at the Bauhaus, it did not take her long to shed the past and leap ahead. This wall hanging, her first at the Weimar Bauhaus, was a radical foray into geometric abstraction. At the same time that Piet Mondrian, in Paris, was making art using nothing but straight vertical and horizontal lines that met at right angles, so was Annelise. Reducing her vocabulary to black and white thread, she gladly discarded the notion of decorative materials, and made the structure and raw components of weaving the source of its beauty. Process was to be celebrated, not disguised, and what she created provided a rich diversion from her everyday concerns.

together with the response: "This is not a course in painting technique; it is a course in imagery." (Given the importance of pixels to Steve Jobs, it seems likely that Seurat was a significant artist for him.) Albers said, "This I like, boy. Which of those bastards in art history don't you like?" I named them, and he asked, "What does your father do?" I told him my father was a printer. "Good, boy, then you're not just an art historian. Then you know something about something."

Both of the Alberses were interested in my connection with printing. Josef showed me the typeface he had made. His favorite designer of the century was, he said, Jan Tschichold, and his favorite typeface Tschichold's Sabon (at which you are looking right now). Fonts mattered to him; Josef revered them in the same way Steve Jobs did when he told the Aspen audience about the importance of the Lisa as a means of fine-tuning the graphics that had been inadequate in earlier computers.

Anni said she was curious to see the processes of commercial printing done at Fox Press. She had worked at some of the greatest art printshops, using lithography and screenprinting and etching in unprecedented ways, and wondered what she might do with methods used for mass production. When I fetched her to take her to the printing plant, an hour from her house, Josef gallantly accompanied her to the car to see her off. He told me, not for the first time, how much he admired the hatchback design of my MGB GT. The carpeted storage area behind the two seats was, he said, ideal for the safe transport of an *Homage to the Square*. The two-seater car was a good example of "nothing wasted, nothing bigger than it has to be."

When we got to the printing plant, built in the late 1950s, Anni said she liked its simple industrial style. In her tan suede jacket and beige skirt and white blouse and tan suede shoes,

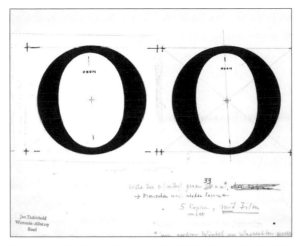

The great Czech designer Jan Tschichold was the best in his field for Josef Albers because of the grace and clarity of his work. Tschichold used the art of modernist painters—Josef among them—to demonstrate the rhythm and sense of proportion that could be applied to graphic design. He became the main designer for Penguin Books, his work having vast influence. These are sketches he made for Sabon, a typeface Josef deemed so elegant and legible that it was eventually carved into his and Anni's gravestones.

using a cane but with impressive posture in spite of an infirmity of her legs, she exuded dignity and composure as she walked in. Then, entering the building, I committed what I now consider to be a conversational sin. I told Anni that when he was building Fox Press, my father had the opportunity to buy a large sculpture by David Smith. Called *Standing Lithographer*, it had a steel type case as the man's chest. "It was priced at only ten thousand dollars back then. Then the architect realized that a fire door was required between the plant and the offices. The fire door cost ten thousand dollars. So the money that Dad was going to use to buy the sculpture had to go for the fire door instead."

As we walked into the offset shop, Anni grimaced. She pointed with her right hand at a large, two-color printing press, recently arrived from Zurich. It was rolling, with sheets of colorful glossy paper stacking up at the end, the smell of ink still strong. "You see that machine," Anni said. "It is far more beautiful than anything David Smith ever touched."

The iPhone is comparable. The components are synchronized to work in tandem and facilitate useful action. And tales of woe about "we could have bought the Picasso for only . . ." are pointless. Worse still, to people with the Alberses' style and values, stories that emphasize the rise in prices of an artwork rather than its more lasting qualities are coarse.

7.

Anni Albers had an unusual connection to the telephone as an object. She rarely referred to the wealth of her natal family, whose conspicuous lifestyle embarrassed her. But her mother was a member of the Ullstein family, rich publishers. One day, Anni recalled with a laugh, her maternal grandfather was given the second telephone in Berlin. When told how to pick up the receiver when it rang, he replied, "Bells are only for domestics."

The world in which the Ullsteins had special clothing made for opera performances and owned planes that delivered magazines to foreign cities had, of course, been shattered. And Anni had married Josef. Differences in social class and family wealth were unimportant to these intrepid people. Anni and Josef used telephones the way that everyone else they knew did, and marveled at the capacity to hear a voice from New York or Chicago through the receiver while they sat there in Orange, Connecticut. Besides, Anni's own mother ended up almost as "a domestic" herself. After being lucky enough to flee Nazi Germany in 1940, Toni Ullstein Fleischmann was reduced to waiting tables. In her former life, she merely rang a bell in the dining room to have the butler or cook come in. But she was still alive!

While iPhones themselves have become expensive, other smart-

A simple four-page manifesto summoned students from all over when the Bauhaus was founded in 1919. To illustrate his succinct text advocating the joining of "craftsmen and artists" stripped of "class pretensions" to create "a clear symbol of a new belief to come," Gropius commissioned the German-American artist Lyonel Feininger to make a print of a Gothic cathedral, the embodiment of a joint effort of masons and glassmakers and skillful people from every discipline to make a masterpiece useful for all of humankind. Feininger's abstracted version of a cathedral soaring heavenward put the medieval building type in modern form. To Gropius, places like Chartres, and the Duomo in Milan, lifted the ideals he put forth in this text, where he writes: "There is no essential difference between the artist and the craftsman." He wanted the Bauhaus's triumph to come from "the hands of a million workers like the crystal symbol of a new faith."

phones are affordable to people even if they don't have the electricity in their homes needed to charge them. Transcending issues of class or personal wealth was an imperative of Bauhaus design.

8.

Two years before Walter Gropius wrote the short manifesto beckoning painters, weavers, woodworkers, silversmiths, masons, typographers, and others to the school he named *the Bauhaus,* he had been the sole person in his army troop to survive a bomb blast that reduced a building to rubble and crushed everyone else inside. Prior to the world war that had beckoned him into the army, Gropius had been a successful architect in Berlin. He had broken precedent in the crisp and clean forms in his pioneering Fagus Factory. Its tensile steel struts, sheer planes of glass, and handsome cantilevers interacted with musical rhythm.

Gropius was an artistic radical, but he was born in a conservative milieu where tradition reigned. He belonged to the social class where he was expected not just to be a soldier, but to be a stylish one. His aristocratic forebears included military heroes and accomplished equestrians as well as the highly touted architect Martin Gropius. The role of the elegant women in the family was to accommodate their husbands' needs and organize domestic life with style.

The young Walter Gropius had had no problem running up substantial expenses at the boot maker's and saddler's and tailor's. He enjoyed his privilege and good looks, and his personal charm guaranteed him lots of fun. But when duty called, he was tough as nails.

After that bomb fell on the structure sheltering Walter Gropius's regiment in the full furor of the First World War, he carved out an air shaft through the crumbling mass overhead. He managed to breathe through the small hole for two days until he was rescued. While their backgrounds had nothing in common, the man who started the Bauhaus and Steve Jobs were alike in their sheer tenacity. These were individuals who never halted at obstacles. Possessed by their goals, they always vaulted all hurdles, personal or professional. And they deliberately allied themselves with other bright and talented people who helped them achieve their dreams.

A year after Gropius survived those two days underneath concrete and bricks and wood, the Grand Duke of Saxony had summoned him to discuss the prospect of a new art school that would have its headquarters in the academy where both fine art and the practical arts had been taught before the war in Weimar. The small German city had become a cultural capital when Goethe and Schiller had lived there a century earlier. The swashbuckling Berliner accepted the offer, and set out to build up a base of supporters and to attract students and teachers.

Gropius was a charismatic persuader of others. His and Steve Jobs's style and speaking manners could hardly have been more opposite—the relative formality of a Berlin aristocrat as opposed to the deliberate casualness of the son of the modest mechanic and schoolteacher who had adopted him—but when Gropius explained Bauhaus values, he had, like Jobs, the conviction and the sense of service to humankind to inspire his audience.

In his private life, Gropius was as boundlessly energetic as in his professional endeavors. With his movie-star features and irresistible smile, he made an immediate impact on most women. He was worldly enough to know how to wear his spats without thinking about it, and carried his soldier's body with an apparent lack of self-awareness. When he was at a fancy sanitarium in the mountains before the war, being treated for one of those maladies that afflict characters in Thomas Mann novels, he had encountered one of the great femmes fatales of all times, Alma Mahler. Mahler was there because of a dark depression following the death of a young child whose father was her much older husband, the renowned composer and conductor Gustav Mahler. The doctors had tried every form of treatment for her—exercise regimes and special diets and mineral baths—and, as a last resort, proposed that she dance with a handsome young man to improve her mental state. Gropius was just the ticket. In this mountain enclave, where most of the men were out-of-shape, pretentious snobs, he was enthusiastic, imaginative, and physically fit. Happily lusty, he had what it took to become Alma's lover on the evening after their first waltz.

The only possible hitch was that, aside from her being married, Alma was traveling with her remaining child and her mother. But her mother obligingly abetted the love affair by taking care of the little girl and writing Gustav that the reason his wife was not often in touch was because she was so busy with her cures. Alma's mother was content to see her beleaguered daughter return to joy through the most effective cure of all— lovemaking with square-jawed Walter Gropius.

People from opposite milieus can have a lot in common at their cores. Steven Paul Jobs was, above all, another extraordinary human being who transcended his origins. Besides their both being undaunted by any notion of the impossible, Jobs and Gropius shared a will to perfection with objectives that mattered more to them than anything in their personal lives, however compelling their private passions were. Gropius's conquest of Alma pleased him, but it would matter more to take the

Bauhaus vision into the stratosphere. Jobs was composed of the same stuff. He would settle for nothing less than having a unique impact on the larger world, and, like Gropius, recognized a rigorous design aesthetic as the route.

9.

Neither Gropius nor Jobs adhered to societal expectations about love any more than about work, and both could make what we think of as major missteps. Gropius could seduce with panache, but he could also mess up. After Alma Mahler left the sanitarium once her mother could no longer justify her absence from her home and husband, he could not bear the separation. He began to plan trysts in Vienna. He developed a scheme for the lovers to stay in different hotels, using pseudonyms. He excitedly wrote a letter to Alma elaborating the details. He asked her to verify that her husband did not have the slightest awareness of his and her affair before finishing the missive with lusty memories that made their ardor and their deception palpable. He folded the letter carefully, put it in an envelope, and sealed it. Then he addressed the letter before posting it. The name he wrote on the envelope for the recipient was not Alma's, but that of Herr Gustav Mahler himself.

The composer handled the situation with aplomb. He summoned Gropius to his and Alma's home. Gustav calmly gave Alma the choice: her lover or him.

Alma decided on the man who was in a position to dictate the terms. Besides, they had a child. But it was not long before she strayed again. She not only returned to Gropius's embraces, but took the artist Oskar Kokoschka as her lover as well. Gropius discovered this at a Kokoschka exhibition when he recognized the woman shown passionately embracing the artist himself in a boat in a storm-tossed sea as none other than his own inamorata.

But Alma's power was supreme. Four years after Gustav's death, she married Gropius, and so was his wife when he started the Bauhaus.

While assembling faculty and raising money and beckoning students to create an institution with which he intended to change the way the world looked, Gropius coped with his demanding wife and, invariably, at least one other woman as well. But the Bauhaus was always center stage, and the sole object of his faithfulness. Ironically, in the mid-1980s, one of the reasons Tina Redse, Steve Jobs's great love at the time, decided to end their relationship was because he was "too influenced by the Bauhaus movement." She accused her lover not just of Walter Gropius's messianic zeal, but of Gropius's same objective: "Steve believed it was our job to teach people aesthetics, to teach people what they should like." According to Redse, the Bauhaus, specifically, provided Jobs with a model for educating the masses in vision and taste.

When he was starting out in his professional life personal relationships mattered less to Jobs than his mission did, as would become all the clearer once his daughter Lisa told her story.

The German architect who came into his own at the start of the twentieth century and the American inventor/businessman who made his mark some sixty years later both cared more about the universal public than the individuals at their sides. So it has often been for political leaders and passionate artists. For better or worse, the mission is paramount.

Part III

1.

The word "Bauhaus" means "house for building." Its lean construction of two rhyming syllables is itself a work of art. This invented term was revolutionary. Since 1908, the institution had been the Grand-Ducal Saxon Academy of Art and the Grand-Ducal School of Arts and Crafts. Then it became legally the "Staatliches Bauhaus in Weimar." But everyone called it just the Bauhaus. The seven-letter sobriquet exemplified the clear design and matter-of-factness seminal to the school's program.

Mies van der Rohe, when he became the third director of the Bauhaus, said the name was Walter Gropius's greatest accomplishment. Mies was damning his fellow architect with faint praise, but he was genuine in his admiration for the spark and shortness of the term. "Bauhaus" also was slightly ambiguous. "iPhone" would be the same: concise, previously nonexistent, snappy, and provocatively inexplicable.

2.

People who invent new words cannot expect them to be taken on board easily. A wordsmith with the audacity to put something new into the vocabulary of his own language, with the expectation that it will take hold in other languages as well, has gone to battle. Steve Jobs knew how important the right words would be to make complicated technology familiar and comfortable. When he started his business for making computers and named this future industrial giant after the piece of fruit most often thrown into a child's lunchbox, it was the first of several occasions when he had to fight for the term he invented.

When Jobs came up with "Apple"—a spectacularly ordinary, almost childish name for a company making instruments that defied most people's understanding—his colleagues protested it. Then, when his colleagues came up with iPod—the original "i" name—Jobs initially rejected it. But the names are central to these success stories. "Bauhaus" gave a cheerful jauntiness to the complex process of building construction. It set the standards of eliminating anything unnecessary. "Apple" is, instantly, a summoning of optimism and pleasure. With shiny skin and a wonderful roundness, it is the perfect piece of fruit. It also suggests a superstar, as in the wonderful child who is "the apple of his parents' eye." And iPhone is, right away, a felicitous breaking of the rules, with the first letter lowercase and the second that capital "P."

But simplicity rarely starts simply. It usually begins with myriad elements converging. The complex and confused heap then gets pared down and reduced. Bach concentrated music to its essence, but he started with a vast range of components only to minimize them to marvelous instrumental solos. "iPhone" also was the end result of a process of revision and winnowing.

It has in common with "Bauhaus" that it never existed before, but derives from known words. "iPhone" has its roots in "telephone" the way that "Bauhaus" has roots in the verb *bauern* and the noun *Haus.* But it was sparklingly new.

"iPhone" is so named because it descends from the family that first used the lowercase *i* before an uppercase letter with the "iMac." Apple Computers had, shortly after the company began, produced a product it named "Macintosh"—presumably because a "Macintosh" is a well-known type of apple. Steve Jobs then wanted to name a new computer model "MacMan." Ken Segall, the executive at Chiat\Day, the Los Angeles advertising agency responsible for the Apple account, was appalled. Jobs's "great name" of "MacMan" for the new Macintosh computer would, Segall maintained, "curdle your blood." Segall proposed the name 'iMac" instead.

Segall said that the *i* was for "Internet." But the ambiguity, and the charm of its being lowercase in front of the uppercase letter that by tradition starts a proper name, are essential to its appeal. And the *i* also has a range of possible associations, including "intelligence," "individuality," "innovation," "imagination," "inventiveness," and "identity." Its elusiveness is part of its power.

Jobs initially rejected "iMac." There was pleading followed by counterproposals, only to have Jobs refuse "iMac" even more vehemently on the second go. He had longer names in mind; more significantly, like the top tier of artists at the Bauhaus, he had an innate resistance to the suggestions of others.

3.

Ken Segall would recall the initial presentation, and subsequently the pivotal naming of that first "i" device. It was in 1998. Jobs had summoned the Apple team to see a new computer in Cupertino. Jobs had already let it be known that there was a product that would save the company, which was in trouble. It was called "c-I"; it was not yet to have another name.

When everyone was assembled, they saw a gray sheet covering a lump on a table. The *c* of "c-I," which one might have assumed was for "computer," stood instead for "consumer." A product manager lifted the sheet, and "the group let out a collective 'holy cow' . . . because it shattered every idea of what computers were supposed to look like. It was a colorful one-piece computer which showed off its inner-circuitry through a semi-transparent shell."

After everyone reacted, various other models were presented, but then they came back to "c-I" as the best. This was not the most literate or erudite group of people in the world, but one would have loved to know the reaction when Jobs said, "We already have a name we like a lot, but I want to see if you guys can beat it. The name is 'MacMan.' "

Segall considered "product naming" one of his specialties. "From some companies . . . you see names like 'Casio G'zOne Commando' or the 'Sony DVP-SR200P/B' DVD Player." (No exaggeration—these are real names, he explained.) He wanted the new product to have a name as original and streamlined as it was. He finally prevailed with iMac. Jobs accepted it the third time round, and the rest is history.

4.

The idea of shortening names had become a vogue at the same time that the name "Bauhaus" was becoming known not just in Europe but also in the Americas and Asia. For leading artists, as well as other major modernists of the 1920s, it is as if, in the ebullience of peacetime following the end of World War I, there was a mass movement to the Renaming Offices. Josef Albers had, even before reaching Weimar, changed the spelling of the

"Joseph" originally given to him, deliberately eschewing the illogic of *ph* sounding like *f* and getting straight to the point without any nonsense or possible confusion. The *ph* was gratuitous, after all. When Annelise Elsa Frieda Fleischmann arrived in Weimar in 1922, she cropped her name with the same jaunty decisiveness that she shortened her hair. The first step was to get rid of the two middle components. Then came the trimming of the first name when she married the man who gave her the right to use Albers in 1925. She relished that change of her last name. Albers had none of the slurriness of Fleischmann, let alone its identification of her family as Jewish and its meaning of "butcher"—literally, "meat man."

The name "Jobs" itself seems like an invention, either shortened or greatly transformed from something more complex. Many people assume that Steve or his parents or someone in a previous generation created it from an eastern European name like "Jablonski."

There is, in fact, no record of any of Paul Jobs's ancestors ever having changed the name from anything else. The name Jobs has been around a long time just as it is. One theory is that it descends from Job—as in the Old Testament character who had so much patience. Other possible origins are "*job*," a French adjective that translates as "unfortunate," or "*jobbe*," the word in Old English (the English in which *Beowulf* was written) for a four-gallon copper vessel. It may, on the other hand, come from "*jube*," a term in medieval French meaning both a long woolen garment and the makers of those long woolen garments.

Ironically, the name "Jobs" sounds as if, like "Bauhaus," it *was* invented—with the goal of being compact and modern. Steven Paul Jobs, in choosing to be known professionally without his full first and middle names, completed the package. The name "Steve Jobs" is both concise and informal. You would not want the creator of the iPhone to have a name that was anything more than one syllable for the first name and one for the last.

The name "Steve Jobs" has an irresistible coherence with the products he invented. And "jobs" is just perfect when associated with an instrument that has a daunting capacity to perform as many tasks as the iPhone can. It also feels synthetic, like the iPhone itself.

A name serves as casing. The Bauhaus placed a premium on the quality of surfaces: the sheen of chrome, the transparency of glass, the declarative hardness and resistance to blemishes of stainless steel, the playfulness and rich texture in offbeat mixtures like the combination of rayon and hemp and silk for woven textiles. The iPhone's physical case is, as we have seen, fundamental to its appeal; the name is also vital.

What Segall—always "Ken," although presumably "Kenneth" on his birth certificate—gave Apple's products adds to their spark. The *i* is cheerful, modest and without ego because it is not capitalized. With minimal energy, it makes a pivotal difference to the word that follows—whether it is "phone," "pad," "pod," or "tunes." Those other words on their own are old hat. It is the *i* that declares them different, younger, more efficient. How concise, how direct, how brilliantly simple—how Bauhaus—to establish the identity of Apple's products with nothing but a single, modest, lowercase vowel! These names are even more compact than the apples one eats. And as poetically modern as "e. e. cummings" or the texts in which Gertrude Stein throws the rules of capital letters to the wind.

5.

Among the modernists who turned their original names into something more suitable to their aesthetics, Charles-Édouard Jeanneret made himself "Le Corbusier." (At least he did so some of the time; he kept his original name as a painter while simultaneously using the new one as an architect, except when he violated his own rules, which is something he loved to do.) People believed at the time he came up with "Le Corbusier" that it was adapted from the "Courvoisier" of his only fancy ancestors. But there are many more reasons that it suited the man who, for one thing, wanted to be perceived as French, not Swiss. "Le Corbusier" sounds like a machine that encompasses a lot—a powerful machine. It also resembles the French word for "crow"—"*corbeau*"—a distinctly unpretty bird to which Le Corbusier often likened himself. Conveniently, it could be shortened to "Le C"—which is how the French often abbreviate "Le Christus"—ideal for the man who saw himself as a messiah and a martyr, in both the value of his mission and the intensity of his suffering. This self-proclaimed "Le C." had marked up his copy of Ernest Renan's biography of Jesus wherever Jesus had the attributes that most enthralled him.

It isn't always that names were shortened; sometimes they were simply changed for effect. In 1919, Frédéric-Louis Sauser felt bogged down by his hyphenated first name and undynamic family name to become Blaise Cendrars—a name that evokes fire and ashes. For a sharp-tongued, playful poet happy to startle his readers, it was a worthwhile switch. At about the same date, the Belgian Fernand-Louis Berckelaers became Michel Seuphor. "Seuphor" had all the same letters as "Orpheus," reconfigured. He made his reputation as a painter, playwright, and art critic, happy to associate his name with the great musician of ancient

Greece, however obliquely. The issue of truth was not all that important to him anyway; he wrote an error-laden biography of the painter Piet Mondrian and, toward the end of his life, had a role in the selling of some forgeries of Mondrian's work.

Pieter Cornelis Mondriaan became quite simply Piet Mondrian when he stopped painting windmills and lighthouses and started to develop clean geometric art. It is, however, erroneous to say—as do almost all the books to date that mention the subject—that he did this when and because he moved to Paris, or even of his own free will. Mondrian made the switch because his uncle, Frits Mondriaan, a well-known painter in Holland and initially Pieter's mentor, so hated the transformation of his nephew's artistic style that he insisted on the dropping of the second *a* to prevent their being confused in the press.

About twenty-five years ago I interviewed Jil Sander for an article for a short-lived magazine called *Benefactor*. I assume Sander needs no further identification; even if, in our peculiar era, the names of most great artists, beyond the household ones like Picasso and Rembrandt, require amplification, commercially successful fashion designers are well known to the general public. My subject was these couturiers as arts patrons. As we sat there in her office in Hamburg while the beguiling purveyor of very expensive and very simple, refined clothes answered my questions, I realized that her name is as trim and lean as the black jackets and gray trousers in which she put both women and men. Moreover, although "Jil" is clearly a woman's name, it was hardly a flowery one along the lines of, say, "Rosemarie" or "Marianne," more usual among German women of her generation. At the end of our discussion, I told her that she and her attitude reminded me slightly of Anni Albers—another fiercely independent woman determined to defy expectations of her role in the world and simplify materials and the shape of things. I went from there to saying that I assumed that, like Anni's, her name was part of the same will to invent and modernize herself. Since her name was as lean and trim as her clothing, she presumably had crafted it much as she had streamlined jackets and trousers. Her public relations majordomo, standing behind her when I asked the question, looked as if she wanted to pull out a pistol and shoot me. Jil Sander, who resembled a healthy, well-weathered ski instructress, a very beautiful one, as natural as she was robust and pretty, broke out into a broad smile and responded to the question even as she knew that the henchwoman behind her was seething. The designer known as "the Queen of Less" did not equivocate in saying, with rolling eyes, "Heidemarie Jiline Sander," pronouncing the S of "Sander" with a rolling Z.

Name changing is sometimes a means of disguising being Jewish. Epsteins become Eastmans, Rosenbergs evolve into Randalls, Greenbergs are suddenly Grays.

The architect Frank Gehry rationalized such a change with an extraordinary justification. "Gehry" was "Goldberg" before Frank changed it. His childhood was marked by both the values of many Eastern European Jewish immigrants to the New World and a keen awareness of the anti-Semitism that prevailed in the society where he hoped to flourish. His upbringing in a close family that emphasized hard work and Talmudic knowledge is the familiar stuff of many life stories. But his youth in the Jewish quarter of Toronto had some remarkable quirks. Frank's grandfather gave him the role of "Shabbos goy"—the non-Jew who served observant Jewish families by performing all the tasks forbidden on the Sabbath. Rather than resent it, the future architect enjoyed running the family hardware store while the others kept observant; that energy and sanctioned irreverence would stay with him forever after.

"Nothing was sacred," Gehry told his biographer, Paul Goldberger. Having been a keen student of the Torah who aced his bar mitzvah reading, he quickly became a teenage atheist, sought out non-Jewish friends, and went his own way.

When Goldberger describes how Goldberg became Gehry, he portrays a blend of someone who manipulates and is manipulated. Gehry says he opposed the name change. "I didn't want to do it. You have to understand, I was super lefty, I was involved with liberal causes," Goldberger quotes him saying. Gehry felt it was "a cop-out." But Anita, the woman he had recently married, insisted. She was Jewish, too, but her maiden name of "Snyder" had been less telling, and she had no use for "Goldberg." Frank's mother sided with Anita; his father objected. Frank, declaring his wife "one tough operator," tells Goldberger: "If you knew Anita, you knew that I had to do it. I had no way out. I was in a corner." He designed his new name architecturally. He invented "Gehry" to simulate "Goldberg" by starting with the same large, circular G, then maintaining the dip to a small, lower-case rounded vowel, then using a high-profile, skyscraper-style letter in the middle, next going small again with the same "r" as in Goldberg, and ending with a tail that slopes inward bracketing the whole by mirroring the opening curve that encases it from the other direction.

When Lincoln Kirstein, the ballet impresario and art and photography connoisseur, one of the most effective and tenacious patrons of modernism in the last century, led people to believe that the "Lincoln" was for Abraham Lincoln, he left out the details. He was named for Abraham Lincoln Filene, always called "Lincoln," son of the founder of Filene's, a department store of which Kirstein's father was chairman. "Filene" itself was a name invented at Ellis Island. So many immigrants named Katz had arrived one day that the man writing the names down as people filed off ships from Europe declared, "Enough Cats! I'll put this one down as 'Feline.'" The future retailer subsequently transposed the *i* and the *e* in his last name. Lincoln Kirstein was accurate in claiming America's sixteenth president as the direct source of his first name, even if his grandparents had used it for their son to honor a hero of their new country rather than because of genetic ancestry. But the founder of the New York City Ballet would never refer to the less glorious relevance of cats to his nomenclature.

6.

"iPhone," like "Bauhaus," would, as a name, be new and provocative—even if it was the direct descendant of Apple's other products whose names started with lowercase *i*'s. California was a place well suited to it. Here, it was not unusual to start from scratch. A William Hamilton cartoon that appeared in *The New Yorker* in 1981 shows two quintessential well-heeled New Yorkers— she with her hair in a bun, he sporting repp stripe tie—looking at a traditional oval-shaped portrait in an ornate frame, with the caption "My goodness, I had no idea people from California had ancestors!" This was the place without a lot of rules and shibboleths, where the public was less likely to question an unprecedented name.

Like "iPad" and "iPod," which preceded it, "iPhone" is a declaration of rule breaking and starting fresh. And, like "Bauhaus," it humors us with its punch and rhythm. These qualities make "Bauhaus" and "iPhone" so strong that they never need translation, and keep their identity unchanged in all languages. Of course they are not the only words that do so. "Singer" is generic for sewing machines worldwide, "Hoover" for vacuum cleaners, "Kleenex" for facial tissues. But those first two cases are people's names, not an unprecedented composition of letters. The invented names have greater zing. At Yale Law School, a renowned anthropologist from Warsaw, instructing students in a lecture hall that he had made copies of a text he wanted them to read, used the word "Xeroxed," only to ask the class the question "'Xeroxed'? 'Xerox'? Do you have those words in English as we do in Polish?"

The obliqueness of the invented name empowers it.

7.

Bauhaus products, like the iPhone, have a look of optimism, of solutions en route. The same is true of the best paintings to emerge from the Bauhaus. The glorious rhythmic and musical abstractions of Paul Klee; the expansive compositions of Wassily Kandinsky; the wall hangings of Anni Albers that, while woven, function like abstract paintings: all are celebrations. There is no visible evidence of struggle, no bitterness, no breast-beating about one's woes, but, rather, exultation, graceful movements, and the sense of ongoing wonder and possibilities.

For people like Gropius and Jobs to create objects that manifested an unruffled state of being and suggested only great things to come, they had to appear in the same mode as individuals. It was vital to put up with the inevitable vitriol of others, not to be thrown by competitiveness, and to set aside the envy and machinations of others.

Walter Gropius certainly had to be thick-skinned with Mies

van der Rohe, as with lots of other difficult people and issues in life. The quip about the name "Bauhaus" being his greatest achievement implicitly disparaged all of Gropius's architecture and minimized the act of actually getting the new school up and running. Putdowns of this sort were the norm in Gropius's life, but the relationship to Mies was particularly fraught, Mies, the scrappy kid from a poor family in Aachen who more than once found himself dealing with the police after his latest pranks or fisticuffs, was jealous of Gropius's aristocratic roots and of the Berliner's privilege of starting at the top. (Georgia van der Rohe, one of Mies's three daughters, was vocal about her father's aristocratic yearnings as well as his rejection of her mother, who came from a wealthy family whose friends were his fashionable clients when Mies himself was a penniless architect. Georgia resented his use of her mother's position and was equally bitter about the austere, man-tailored dresses her father's mistress, the designer Lilly Reich, forced her to wear when she would have preferred floral patterns and a bit of lace.) Power plays, jealousy, and backbiting figured in life at the Bauhaus in each of its incarnations, as they did in the history of iPhones and iPods and iPads and other Apple products. People like Gropius and Jobs had to have the ability not to be bogged down by any of those negative diversions from their purposes in life.

Steve Jobs would have to avoid crumbling from competitive people trying to top, and topple, him, both within his own empire and outside it. The history of Apple—of Jobs's being forced out of the company and, a decade later, having its floundering directors do handstands to get him back in—is well documented. Such behavior is more expected in the milieu of giant corporations than in small art schools, but, in both his case and Gropius's, other people's greed for control and consequent manipulations required both alertness and resilience on the part of the leaders. They had to withstand attacks both brazen and clandestine, never losing faith in themselves and their beliefs. In the modern era, Barack Obama has possibly been the greatest exemplar of that form of strength. These tenacious people at the helm rarely showed strife; all wore faces of optimism and of confidence without arrogance.

Jobs and Gropius would also have to get used to knocking on doors for financial support and to preaching their beliefs to people who might help them reach their goals. And then they would produce things that showed none of the struggle that went into their realization. They eschewed any hint of autobiography in their creations. Rather, they produced designs intended to suggest smooth functioning devoid of impediments.

Beyond that, the products of the Bauhaus and subsequently of Apple give their users a lift. The iPhone—sleek, elegant, and stylish—provides its owner with a feeling of being urbane and sophisticated. The same is true of the "Wardrobe for a Gentleman" made of cherrywood and maple with nickel fittings by Herbert Bayer in the cabinetmaking workshop in Dessau in 1927. Even if the person using it dresses only in filthy jeans and shirts with holes in them, he assumes a certain élan and sense of class just by having his clothes in this stately cupboard. To create objects of such flawless grace, the people who developed them had to learn how to screen out the negative forces coming at them from all directions and immerse themselves totally in their impeccable making. The designs always looked new and ageless; to fabricate them that way, their engineers needed to know how to focus only on Plato's visible perfection.

Part IV

1.

Steve Jobs had started out more as a geek who liked mechanics than as a design aficionado. He was the sort of pale, indoors boy who spends hours tinkering with his ham radio when most guys his age are playing outside. Apple I was born in the Jobs family's garage; Steve stored parts throughout the house. The device he put together consisted of a computer board, a keyboard, and a power supply. There was neither a monitor nor a case. One of many new information devices being developed at the time, it sold, but only to people who were already experts in that sort of gadgetry. The purchasers were knowledgeable hobbyists who understood how to make it run.

Jobs soon realized that personal computers should be single objects in handsome cases, with built-in keyboards. And they should appeal aesthetically. His partner in developing the new tool, Ron Wayne, designed one. It had a Plexiglas cover that sat perched over its gray metal body. Jobs did not like it, although he did not know why. He had no clear vision of what he wanted, but this was not it. Jobs's insistence in rejecting new names, new approaches, and new designs was almost as vital to him as his capacity to try the unprecedented. And when he dismissed something, there was no point in arguing.

That absolute conviction about distastes, and the resolve to stick to your guns after you said no to the second-rate, had been essential at Jobs's beloved Bauhaus. Discernment was as vital as the will to develop new, out-of-the-box possibilities.

The goal at the Bauhaus was to make the implements of everyday life not just more effective but also handsome in an unassuming way. They should facilitate human usage naturally while being clean to look at. That purpose meant that what seemed deceptively ordinary could be a source of inspiration. This was rare in a world where most people wanted their designs either fancy and expensive-looking or seemingly original.

Novelty and pretension sold best, but what counted more at the Bauhaus was an appreciation of what was basic and inexpensive.

Steve Jobs took that cue. One day when he was studying kitchen appliances at Macy's, he saw a Cuisinart. Its molded plastic case was gold to him. The sanitary white of sterile hospital equipment, those food processors for everyday use are made in the hue that creates a neutral setting for action. The color does not intrude or impose. The glistening plastic evokes faith and possibilities. The overall appearance functions like the starting line of a race.

Jobs quickly had a computer case designed that was similar to that of the Cuisinart. Its form served its purpose. The durable white plastic imparted a feeling of ease.

Jobs also cleaned up the circuit board inside the case. Fastidiousness as well as a drive for clarity were further links between him and the Bauhaus Masters—as the school's teachers were called—although they would have found the way Jobs dressed intolerable. The Bauhaus did not do goofy. or slovenly. But the people who developed and perpetuated the standards of the

Bauhaus would have embraced Jobs's goals for pleasing, user-friendly objects.

2.

That the Cuisinart was the parent of the iMac positions the iPhone in the subsequent generation of the family tree of direct descendants from the Bauhaus. The Cuisinart is a multitask food processor that was created by Carl Sontheimer, an American who was raised in France in the 1920s. Sontheimer had left the United States as a baby but had returned to study engineering at MIT. He developed not only this kitchen appliance that has become a staple of American households, but also invented a direction finder, dependent on microwaves, that enabled NASA to land a satellite on the moon.

One day when Steve Jobs was studying kitchen appliances at his local Macy's, he saw a Cuisinart. Its molded white plastic case suddenly struck him as the perfect material for the shells of personal computers. The food processor that had become a staple of American households, but could be traced back to German design advances under Bauhaus influence, led Jobs to make his products more durable and fastidious, evoking optimism and a sense of possibilities.

Sontheimer acquired substantial wealth through the electronics companies that produced these two items. Still, the Cuisinart, unlike the direction finder for space vehicles, was an adaptation, not an original creation. What Sontheimer put on the market in America in 1973 was essentially derivative of the Robot-Coupe, a higher-priced machine invented by a Frenchman, Pierre Verdon.

With its name conjuring a robot able to cut and chop, the Robot-Coupe, which is still made today, is a heavier, more expensive version of the device, industrial in strength, mainly used in restaurant kitchens. The Cuisinart's success is the result of its being sufficiently low-priced, and produced in such quantity, that many households all over America can afford to have one in their kitchens. It and its vast range of clones realize the Bauhaus ideal of good design for everyone. To be useful to gourmet chefs with company money to spend was an achievement, but to lighten the burdens of life for your median-level homemaker was a triumph.

The Robot-Coupe, meanwhile, was itself a form of clone. Just after World War II, the German company Electrostar had created Starmix, a kitchen device that not only chopped and blended and grated, but also had attachments for making ice cream and slicing bread. Its motor could even be used to drive a vacuum cleaner. The developer of the Starmix was Albrecht Graf von Schlitz genannt von Goertz von Wrisberg. This nobleman, born in Lower Saxony, developed his approach to design under the influence of the neighboring Bauhaus. Like so many Bauhaus people, in spite of his more than acceptable background, he left Germany in the mid-1930s and went to the United States. Having no money, and now simply Albrecht Goertz, he first worked washing cars—his obsession. In 1938, he rented a garage near

Los Angeles and began rebuilding cars while modifying their designs. His two-door coupe, the "Paragon," was shown at the 1939 World's Fair in New York.

Goertz served in the US Army during World War II. When the war ended, he met the designer/architect Raymond Loewy. Loewy, who had created the Studebaker, admired the sporty Paragon Goertz drove. In little time, Goertz designed his own Studebaker, a BMW sports car, and the Starmix.

When Steve Jobs took his inspiration from the Cuisinart he saw at Macy's, he effectively started a new generation of the family of industrial designers formed by the Bauhaus.

Jobs's inspiration in the appliance section of a department store was in the spirit of Marcel Breuer determining an essential component of his furniture by studying bicycles. Breuer recognized in the hollow chrome tubing of bicycle handlebars the ideal material for the framework of his armchairs. Anni Albers found the right drinking glasses to use at her and Josef's first home, at the Dessau Bauhaus—having failed in every level of emporium for tableware, high and low—when she finally satisfied her insistence on clear, graceful no-nonsense design by inadvertently discovering the glass beakers used in chemistry labs. In Connecticut, half a century later, Anni battled similarly to obtain a simple lighting fixture to mount over the staircase of her and Josef's raised ranch house. She ended up adapting one meant for outdoor use. Spotting her quarry within it, she removed the garland of metal flowers encasing the plain glass cylinder the decoration was intended to conceal. Usually frugal, for once Anni happily made an exception by spending too much only to throw out the parts of the fixture that had driven up the cost.

In the kitchen, Anni Albers used a three-tier white metal rolling tray made for use in a hospital. Anything intended "for home use" was "too designed." Her bedside table was a typewriter stand. By most standards, this brown metal object was ugly and somehow unfriendly looking. But it suited Anni because you could adjust the height and it had sufficient surface space for her small Sony television as well as the telephone, box of Kleenex, writing implements, and sketchbook she wanted in easy reach at all times. Where other people would have had chests of drawers in their bedroom, Anni had office storage cabinets faced in brown Formica.

When they first moved to New Haven, the Alberses had old car seats, removed from a wreck, as their living-room sofas. This was the same offbeat approach that guided Steve Jobs when he was considering the design of a high-tech instrument. You needed to be willing to look in one domain for inspiration in another. You had to forget tradition, even affront it. What counted was both efficiency and the capacity to please the user.

3.

Once he had developed the design he wanted, Steve Jobs joined the Homebrew Computer Club, of which his friend and colleague Steve Wozniak was a member. Jobs and Wozniak started to build and print circuit boards in the Jobs family garage. They stored spare parts in an empty room in the house. In 1976, Jobs proposed that they start a business, primarily to sell those circuit boards to their fellow members of Homebrew.

It is a history that has been told many times, both in quick summations and with copious details. The difference this time is that you should consider the activity in that garage as extraordinarily similar to the searching and trying that took place in what were called "workshops" at the Weimar Bauhaus and "laboratories" when the school moved to Dessau. The process was among the most wonderful of human endeavors: experimentation. Intelligent, daring individuals were venturing into new territory with only a vague notion of what the results might be, applying knowledge adventurously and letting various materials, some recently developed and others tried and true, realize their possibilities in new ways.

In those days, Jobs was spending a lot of time at All One Farm. He had started going to these two hundred and twenty acres of apple trees when he was a student at Reed College. Located forty miles from Portland, Oregon, All One was a commune, very much in the spirit of the times, with Hare Krishna monks cooking vegetarian food. There Jobs found happy refuge from the academic studies and rigor that did not suit him, and he liked the orchards so much he ended up leading the crew that pruned the Gravenstein apple trees and processed cider.

Jobs had recently returned to All One before flying to meet with Wozniak. This was the moment he came up with the name "Apple Computer," "because it sounded fun, spirited, and not intimidating." To put such a friendly, natural name in front of the word "computer" made the new devices have less of the air of cold, potentially nerve-racking accounting tools. It gave a sense of something homey and familiar.

That reverence for the qualities embodied in apples is central to the Bauhaus/iPhone continuum. Apples themselves are perfect examples of objects in which every minute detail serves a purpose. They are pleasing to look at, and they provide service and enjoyment by being both nourishing and delicious to eat. The symbol Jobs picked was in keeping with Bauhaus ideology: straightforward, devoid of any ornament, purposeful, and appealing to everyone, everywhere. Jobs had made a brilliant choice. Apples exemplify the fully developed fruit that Paul Klee likened to the end result of the creative process, in which the earlier stages are the equivalents of seeds being planted, roots taking hold, trees emerging, and their leaves taking form. With their strong and taut skin containing the miracle within, they exemplify perfect packaging. They provide multitudinous benefits for their consumers. Apples share with the greatest output of the Bauhaus a universality and lack of pretension.

The choice of "Apple" was important in the way that the invention of "iMac" would be. Starting up his new company to produce computers affordable to individual consumers, Jobs recognized the importance of names as part of the packaging.

This was before the horrid term "branding" became ubiquitous, associating the commercial importance of high-status labels with the cruel practice by which the ownership of cattle is burned into the creatures' flesh. "Networking" probably came into use at about the same time. These words have become increasingly popular now that people openly admit their greed for personal advancement. Human beings have always been ambitious, but people were less brazen in previous eras. No one at the Bauhaus would have dreamed that students should take courses in marketing and self-promotion, as they do in art schools today. The goal was the improvement of human life, seen with the perspective that none of us live forever. Neither fame nor personal financial fortune was paramount. At the Bauhaus, the need simply to survive, with enough money for bare necessities, was relevant to an extent that few histories acknowledge, but when Kandinsky was overjoyed that his appointment to the Weimar faculty enabled him to buy his first new pair of shoes in a decade, he never would have considered that, twenty years after his death, top-level jewelers and international art dealers would be courting his widow to get a piece of her fortune.

Jobs's decision to call the company "Apple" was different from the uncharted territory provided by "Bauhaus" and "iPhone." Those names give the surprise of Oz, of the world down Alice's rabbit hole, of landing on a planet one never knew was there. "Apple," on the contrary, made what was potentially confusing seem familiar and comforting.

Still, even if the name "Apple" provided a sense of familiarity, Jobs was emphatic that objects must refresh and awaken us. "When you open the box of an iPhone or iPad, we want that tactile experience to set the tone for how you perceive the product."

That tone needed to be established in advance. When, a few years after starting his company, he was launching Apple II, Jobs engaged a public-relations specialist named Regis McKenna. What had attracted Jobs to McKenna were his magazine ads for the company Intel. The Intel ads were brightly colored images of poker chips and racing cars; Jobs wanted the same panache for Apple II.

When McKenna did not respond to his outreach, Jobs phoned every day. He finally snagged a meeting with the PR hotshot who had initially considered himself too busy for this unknown entrepreneur. McKenna delighted unabashedly in his own importance. His business card said "Regis McKenna, himself." But Jobs made his case. Also a college dropout, Jobs knew that to convince a fellow autodidact required sheer audacity. McKenna accepted his new client, and in little time hired an art director to create a new logo for Apple.

McKenna would be nicknamed "the Silicon Valley Svengali," saying later in his life that the biggest mistake he ever made was accepting a fee for his remuneration rather than the 20 percent of the stock in the new company he had been offered as an alternative. Indeed, "Regis McKenna, himself" would advance Jobs's mission to make Apple computers startling in the way that the masterpieces of Bauhaus design had been.

A large part of that triumph was because for the people Steve Jobs hired to refine the appearance of his products, and to work out their marketing, Bauhaus leanness as well as pure imagination constituted the holy grail. Moreover, by the time Regis McKenna and other gutsy and out-of-the-box thinkers joined the effort to get the new products widely distributed, the world was ready for brazen simplicity. The temerity and gumption of the stalwarts of the Bauhaus had paved the way for the groundswell in acceptance of modern design. Now, with a few firebrands to enlarge the audience for the new, some with expertise in business affairs and others in advanced technology, efficient and good-looking tools integral to current existence would succeed beyond the Bauhauslers' wildest dreams

4.

When the Bauhaus moved to its Dessau headquarters in 1925, Walter Gropius stated afresh the essential purposes of his reborn institution. "The Bauhaus wants to serve in the development of present-day housing, from the simplest household appliances to the finished dwelling. In the conviction that household appliances and furnishings must be rationally related to each other, the Bauhaus is seeking—by systematic practical and theoretical research into formal, technical, and economic fields—to derive the design of an object from its natural functions and relationships." The Bauhaus director pointed out that, just as people no longer wore historical clothes, they were entitled to the most modern objects in their everyday lives. Even if the suits and ties of the male Bauhaus masters, and the dresses of the female ones, now seem quaint and old-fashioned, and certainly traditional, compared to the jeans and sneakers of the Apple crowd, it all was up for reconsideration.

Gropius characterized what was essential: "An object . . . must fulfill its function usefully, be durable, economical, and 'beautiful.'" He was unambiguous that these design standards were not simply desirable; they were essential to humankind. "The creation of standard types for all practical commodities of everyday use is a social necessity." The time had come to eliminate "romantic gloss and wasteful frivolity." The requisites were "simplicity in multiplicity" and "the limitation to characteristic primary forms and colors, readily accessible to everyone."

The Dessau facilities for textile making, carpentry, metal work, and every other realm of artistic production were larger in Weimar. "In these laboratories the Bauhaus wants to train a new kind of collaborator for industry and the crafts, who has an equal command of both technology and form." The prototypes that were to be developed to perfection by hand would, in turn, be produced in large scale, industrially. They would be manufactured in accord with "all the modern, economical methods of standardization (mass production by industry) and by large-scale sales. . . . The Bauhaus fights against the cheap substitute, inferior workmanship, and the dilettantism of the handicrafts, for a new standard of quality work."

Georg Muche, who taught "form" while others taught technique, articulated the relationship of "Fine Art and Industrial Form" in an article published in the first copy of *Bauhaus,* a journal launched following the move to the new headquarters.

"The attempt to penetrate industrial production with the laws of design in accordance with the findings of abstract art has led to the creation of a new style that rejects ornamentation as an old-fashioned mode of expression of past craft cultures." The beauty of industrial objects was seen to come about as the result of "functional considerations and those of technological, economic, and organizational feasibility." By marrying practicality and efficiency with fine appearance, a splendid oneness developed. For centuries, useful objects had been cloaked in decorative elements. Bucolic picnics had been woven into upholstery fabrics, while garlands of angels busied up plaster ceilings. Now, what made something work well was in and of itself the core of its aesthetic quality.

The development of the iPhone depended on related ideals. Bauhaus values were reborn not just in the aesthetics but in the idea of making individuals an integral part of an organization that was a larger whole. The Bauhaus in its second incarnation had "a business organization" to market the new products. The royalties would go to the Bauhaus institutionally. Apple's strategy was to have every single employee own stock in the company, meaning that profits benefited all of them. The Bauhaus system was a form of "work for hire," with no chance of greater take-home pay, while Apple's was an incentive to increase profitability for personal gain, but each strengthened the bond between participants in an undertaking and the overall entity.

Whatever struggles there were in product development, the final designs were seamless. The internecine rifts, the stories of people slamming doors in one another's faces, either in the workshops of Dessau or within Apple's headquarters in Cupertino, California, are nowhere to be seen in the by-products of their cacophony. Bauhaus tableware, the handsome and comfortable armchairs produced in Dessau, Mac computers, and iPhones: their appearance is triumphant. What we see is unruffled. There may be an experience ahead that is not what we hope for—a meal that could have been better cooked, an unpleasant conversation, a nasty e-mail, information that is not as we wish—but the vehicle fortifies and calms us through the sheer intelligence and grace of its design.

5.

To design the iPhone with minimal distraction and maximum concentration, the Apple design team was given a floor at Apple's corporate headquarters in "lockdown" mode. People entering and leaving passed through a sequence of doors, each requiring a special badge. Security cameras facilitated perpetual surveillance.

This clandestine world for the people designing the new phone was called "the Purple Dorm." Work sometimes continued 'round the clock. The inevitable boxes of pizza were brought in not by regular delivery services, but only by authorized personnel with the essential badges. The front door had a sign saying "Fight Club." In the movie with that name, the first rule of the club is, "you do not talk about the fight club." The second rule is, "you do not talk about the fight club."

The British-born designer Jonathan Ive headed the team. He had the overall vision for the look of the iPhone and steered the detailing. "We're talking about perception. We're talking about how you feel about the product, not in a physical sense, but in a perceptual sense." Saying that, Ive could have been paraphrasing Josef Albers's "discrepancy between physical fact and psychic effect." Albers celebrated the way that color and line have no truth of their own; what matters is their perception, which depends on what is adjacent to them. Ive's emphasis on perception, like Albers's, depended on an emotional intelligence, far more exciting than a chemist's formulas. They cultivated human responses. That element of perception being more important than a knowledge of invisible, humdrum facts reflects the same priority Steve Jobs had emphasized in 1983 when he spoke to his audience in Aspen about Aristotle.

The idea of an iPhone being "an iPod with a rotary dial pad on its screen" was considered seriously before being ruled out. To achieve the "wow" he wanted in the iPhone, Ive kept the screen uncluttered. It was not to dominate, but to provide a sense of possibilities. He needed the tranquility, the suggestion of the infinite offered by a calm sea or a desert landscape. Ive compared his final design to the sort of swimming pools that have an invisible edge, where one has the illusion of stroking toward a drop-off into vastness.

The words that Ive used for what he wanted when the display appeared on the screen were "magical" and "surprising." "Magical" is the adjective Josef Albers used for both color interaction and the imaginative configurations of straight lines to conjure impossible forms. "Surprising" was a favorite of Anni Albers. She used it for shimmering gold threads you do not see initially in textiles that are subdued overall; they seem to come from nowhere. She created surprise by drawing what seem to be patterns but end up having no repetition at all. What initially appears systematic is in fact irregular. "Surprising" is also a favorite of the clothing designer Paul Smith. At an exhibition of work by both the Alberses together, he marveled at the sudden bit of orange he had not anticipated in an otherwise black-and-gray weaving by Anni, and then at more orange details he had

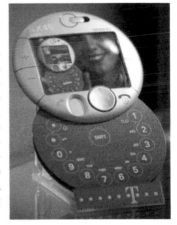

The iPhone is just one type of "smartphone"—a generic term first used in 1995 to describe a range of handheld instruments that served as mobile telephones and personal computers. By the time Apple presented the first iPhone in 2007, various companies were making these objects, but most had designs that resembled jukeboxes and quickly became outdated. When the iPhone appeared, it was like the Barcelona Pavilion compared to the most massive Victorian architecture. Clockwise: Motorola StarTAC Rainbow, 1997; Nokia 9000, 1996; Siemens SX45 (concept), 2002.

not imagined could possibly be there as well. He likened these unexpected sights to his use of an outrageous purple patterned silk lining inside a somber pinstripe suit. No one would have suspected it.

Seeking the "magical" and "surprising" in the iPhone, Ive and his team tried plastic screens and different corners. There was the "Extrudo" and there was the "Sandwich." None of the solutions even remotely satisfied Ive. He had the Bauhaus inability to settle for the second-rate or the not-quite-right. A former Sony designer, Shin Nishibori, now working for Apple, came up with ideas relating to known phones from Asia. Nishibori's mock-ups would eventually be evidence in a courtroom when Samsung sued Apple for copying their designs. Apple won, proving they had never stolen Samsung's ideas.

Ive valued touch as well as appearance. At the Bauhaus, tactile experience had been similarly prized. The difference between the feel of silk and of hemp, the sensation of travertine as opposed to plywood, was always considered. What does it feel like to have so much happen by moving your fingers on a completely smooth, touch-sensitive surface that needs only the lightest contact rather than a keyboard that has to be depressed more muscularly?

When Steve Jobs launched the iPhone at the Macworld Conference on January 9, 2007, he showed the simple object that was smaller than the palm of his hand with an enormous Apple logo projected in snowy white on a black screen behind him. The logo looms godlike, as if to say, "Now there is light." By this point, Jobs had gone casual in public, his simple turtleneck jersey making him "one of the guys" rather than a corporate tycoon: the iPhone, after all, was meant for everyone.

Jonathan Ive, invariably known as Jony, was the main designer of the iPhone. He was educated in England, where his father, Michael Ive, was appointed by the British Education Ministry to make design technology part of the core curriculum of secondary school education. Michael made Bauhaus methodology central to Jony's formation. In this photo taken in 2013, Jony's style is "technology casual," much as Steve Jobs's became. More recently, in magazines like *Vanity Fair*, he has gone upmarket and worldly.

When we shift a car from second to third gear by stepping on the clutch and moving the gearshift up, right, and up again, we have a totally different sense of engagement than when we use automatic transmission. When we drink through a straw, we imbibe a beverage with actions that have little to do with what occurs when we sip it from the side of a drinking glass.

These levels of human experience, and their small yet significant impact on the user, can determine whether an object flourishes long-term or disappears from public use. Jonathan Ive, like the Bauhauslers, evaluated every nuance. And Ive, like the makers of dinnerware at the Bauhaus, delighted in a conscious sense of progress. The pure white porcelain plates and stainless-steel flatware at the Bauhaus were not solely what they were but also a recognizable leap over their heavy, visually encumbered precedents. The embedded keyboard of the iPhone brought with it the feeling of progress over the BlackBerry's, which had a keyboard like a miniature typewriter. To reduce apparent details was to increase emotional and physical ease.

Part V

1.

The iPhone is central to one of the most groundbreaking developments in human history. It does not stand alone, but it shines within the revolution in human communication whereby people almost anywhere on earth can be in touch with others wherever they are. Yet although it is the ultimate realization of the Bauhaus ideal for objects of everyday use, applied to a fantastic new science, it is not "art" in the way that a lot that came out of the Bauhaus is. The distinction has already been made, but, at the same time that we should see the central importance of great design, we should not confuse it with the sort of visual art that transports us to another emotional level and deepens our experience of all of life and provides unrivaled pleasure.

As a tool the iPhone is brilliant, but it is in no way comparable to Paul Klee's paintings from Weimar and Dessau, those homilies to plant growth and the fantastic tropical fish in the underwater universe, to the marvels of the Egyptian desert and to Mediterranean bounty. Nor does it rival Wassily Kandinsky's freewheeling abstractions, so euphoric as to be audible, or Anni Albers's refreshing and salubrious textile pieces with their ceaseless, nonformulaic rhythm that take you away from everyday existence into another sphere. It has the leanness of Mies's architecture, but not the poetry with which its walls and the openings within them become as glorious as a Beethoven duet. The iPhone has remarkable capacities within a neat package of high Bauhaus standards, but it is not the same thing as art that induces poetic rapture like the great achievements of the Italian Renaissance. When we celebrate it, it is as the embodiment of the Bauhaus's push for great design for everyone, but not of the underestimated aspect of that art school: that it provided a haven in which people could make miraculous art.

That's the only caveat. The issue was hotly debated within the Bauhaus itself: whether the priority should be the creation of "studio art" by individual geniuses or the development of design prototypes to be applied universally. It is here that iPhone triumphs.

This second category was the crux of the Bauhaus program for Walter Gropius. Kandinsky, on the other hand, would rather use his antique, traditional Russian samovar than a clean-lined teapot of modern manufacture. Anni and Josef Albers were rare in their passion both for art as art and for understated functional objects. When, as an octogenarian, Anni went to a Picasso retrospective at the Museum of Modern Art, when she was beholding a small classical nude male Picasso had made in the 1920s, she lit up completely. "We got it wrong at the Bauhaus. Man is the most important thing of all," Anni said. But she also marveled at plastic food-storage containers at her local Sears, Roebuck. And saw a connection. The Picasso and the transparent boxes both realized their goals, were technically outstanding, and connected with the essence of being an alive human being—whether that means standing naked in front of the sea or having an efficient object in which to store your leftover beef stew. Josef worshipped

Cézanne, but he was almost as reverent about good cameras. In both he felt the sheer integrity of people doing their absolute best, of creative genius combined with restraint and discipline.

People like the Alberses would have marveled at the extraordinary uses of the iPhone. Of course, Apple's product is just one of many modern devices with similar if not identical capacities, but, with this object that is shorter than the distance from the bottom of your hand to your fingertips, and not as wide as the palm, you can stand in a Brazilian rain forest and know the weather in London or order a special shirt that has built-in sun protection and is made in northern California but will be delivered within two weeks to the hotel you specified in São Paulo. You do all this while looking at a pleasant screen in which the information appears in a thin sans serif typeface, the spacing between the letters millimeter-perfect so as to be legible and pleasing. The lines are ruler-straight. Nothing is crowded, although infinite functions are available.

Tasks are as easy to perform as humanly possible. With an iPhone, four-year-olds can watch the movie *Frozen*. The decibel level can be adjusted so that they can hear it but not disturb other people nearby. The kids can enter a fantasy world in vivid colors while their parents get necessary time to themselves. iPhones make it possible for one business associate to notify another that he is stuck in traffic and will be ten minutes late for lunch; in that capacity, it mitigates potential anxiety and keeps both parties calm. iPhones enable a person cooking clafoutis to consult three different sources on the recommended number of eggs.

A lot of what the iPhone facilitates is simply a matter of access to the internet, available through computers of every form, but its small physical scale and portability enhance the ease of communication. iPhones make it possible for patients to send detailed photos of rashes or bruises to their doctors, who in turn may say they are of no concern or else suggest a treatment; what used to require waiting five hours in a crowded emergency room is resolved lickety-split.

iPhones are also the means through which readings of heart rhythm recorded on a loop monitor installed in a patient's chest can be sent to the technicians for regular review. The patients could be on the North Pole or in sub-Saharan Africa, his cardiologist in New York; the iPhone will facilitate the "all clear" or else call for urgent attention, saving a life. Surely the Bauhauslers would have marveled at the capacity of "Bluetooth" to facilitate communication—even if they might have been puzzled by a term that suggests a grinning pirate.

The appreciation of modern medical advances and focused minds was essential to the spirit of the Bauhaus. And so was the recognition of the human body as one of the greatest creations of all. The heart, whose rhythm is thus monitored, has, like most machines, some parts that wear out with age. Those parallels between manmade tools and living beings were vital to both Steve Jobs and the Bauhauslers. Jobs extolled the natural as a source of the operations executed within his inventions. The Bauhauslers saw in everything from the growth of trees to the functioning of the universe the ultimate creation, and made what they painted or designed echo the processes of nature.

At the Bauhaus, physical harmony was considered a mechanism to induce emotional harmony, to add to balance in life. If a glass tabletop was half a centimeter thick, its support and legs precisely doubled or tripled that measure. The iPhone also depended on the exact relation of proportions, so that a sense of rightness enters the user.

The Bauhauslers themselves could not have imagined it; many people today cannot imagine life without it. And it has become integral to the quotidian on a phenomenal scale that even Gropius and his cohorts at their most grandiose considered unattainable. But the iPhone realizes the qualities that were first evident in Gropius's 1911–12 Fagus Factory, a building of inestimable beauty. Its cantilevered entrance covering, the windows of tensile steel and generous amounts of clear glass that were wrapped around the corners in an unprecedented way, the clean lines playing against one another with Mozartian lightness, were the seeds of the iPhone. The sheen, the precision, the visual simplicity, the whiteness, the reverence for machined materials,

The Fagus Factory was designed by Gropius and Adolf Meyer in 1911. Its generous amounts of clear glass wrapped around the corners, and its tensile steel window dividers, gave it clean lines and a Mozartian lightness unprecedented in industrial design. Its sheen, precision, simplicity, luminosity, and reverence for machined materials were the seeds of the design of the iPhone.

the luminosity, and the clarity that the founder of the Bauhaus brought to perfection in his first major building are the same qualities that Herbert Bayer transmitted in his Aspen building and that are today exemplified by the design of the iPhone. And if you compare the iPhone to an old-fashioned rotary dial phone, heavy physically and visually, you have much the same experience as when you consider the Fagus Factory alongside the buildings that proliferated all over Germany in the nineteenth century, where massive ornamental cornices and marble columns based on ancient Greek prototypes declared the sheer solidity and weight and permanence of the bank or the post office or even the blast furnace. Steve Jobs shared Walter Gropius's courage in developing what is so light as to be almost ethereal, like a billowing, airy meringue in a world as heavy as goulash.

The iPhone, while being only one of the many modern objects that manifest these goals, is the quintessential one. It does so both because of its compactness and effectiveness and because of its incorporation of recent technological advances. It celebrates—and this was something vital to Bauhaus philosophy—the advances of its own time, and eschews nostalgia or historical reference.

In the development of production of the iPhone as in the thinking and output of the Bauhaus, the romance of the "old-fashioned" was not just left behind; it was shunned. So was the idea of materials affordable only to the rich and privileged. So was ornament, deemed gratuitous. But inventiveness and the courage to pursue what had never existed before were prized, as were usefulness and pleasure.

Once it enters your everyday life, the romance, alas, is over. As soon as you deal with service issues and try to get different programs to deliver on their promises, you may feel, as many of us do, like outsiders kept apart from a club of confident "experts" who seem sadistically determined to waste our time and test our patience. But in its initial aesthetics and functioning as an object itself, the iPhone epitomized the highest achievement of the new technology for its capacity to make so much accessible to so many.

2.

The subsuming goal of the Bauhaus was to lighten life's burdens and facilitate an appreciation of its wonders. Tom Wolfe's *From Bauhaus to Our House,* which some of you may know, is a nasty book of tragic consequence because it claims the opposite and lured a large audience. Its snappy and smart-alecky style and facade of authoritativeness blinded readers to the falseness of its premise. Wolfe belittles what in fact was brilliance and mocks what was genuine and heartfelt. Because glibness has its audience, Wolfe's mean narrative undermined the general understanding of the Bauhaus.

To recognize the way that the iPhone does justice to the splendid vision of the community of brave souls at the Bauhaus, you need first to dismiss the attitude that Tom Wolfe inculcated far and wide. What has gone "from Bauhaus to our house" is neither cold nor impractical nor didactic. It belongs to a vision of human betterment.

3.

Anni and Josef Albers's house was the whitest place imaginable.

Outside, this gangly "raised ranch" was covered with shingles that looked completely artificial, a beige between the colors of Band-Aids and of iced coffee with too much milk in it. Bauhauslers believed in synthetics and plastics and all types of manmade materials, and those shingles were an artificial wood composite resistant to the bad weather and temperature changes that are a reality in Connecticut.

The benefits of manufacture tailored for maximum performance of purpose was essential. Natural substances were often too unpredictable. And to be true to the Bauhaus, one had to forget the style of the Bauhaus. Josef was outraged that his former colleague Marcel Breuer, who also ended up in Connecticut, built houses with flat roofs there. They leaked in the winter, of course. What was elegant in Dessau and served as a terrace should not have been expected to work where large amounts of snow melt into thick ice that in turn melts into water. Adapting to current circumstances was a core value of the Bauhaus, and mattered far more than "a look." "German," "Jewish," "Hungarian," "rich," "poor": none of those ways of distinguishing background counted. You would have thought, were it not for his accent, that Josef himself had been born in New England. Wallace Stevens, who lived in Hartford, the capital of Connecticut, wrote, "The thrift and frugality of the Connecticut Yankee . . . were imposed on him by the character of the natural world . . . in which he came to live, which has not changed." "Came to live" was a seminal issue. Mobility in life is expected in the true practitioners of Bauhaus values as in the people for whom iPods are made. Identity by birthplace or nationality makes no difference. Stevens allows that while he was not born in Connecticut, that happenstance made no difference to his sense of belonging. Good design, similarly, works equally well for anyone from everywhere; it obliterates imposed notions of human differences. Stevens was like both Anni and Josef Albers in writing, "It is not that I am a native, but I feel like one. All of us together constitute the existing community."

Stevens writes of people worldwide who "share in common an origin of hardihood, good faith and good will." You felt this initially when you saw the Alberses' house from outside and even more after going in. It was American-style in the way that Betty Crocker cake mixes and the houses in 1950s sitcoms are, but beyond the elements that made it native, it was marked above all by its airy austerity and a leanness so Spartan as to astonish you. That monastic simplicity took one out of place and time.

When Anni arrived in Weimar in 1922, she lived in a single, bare-bones room. She was allowed a bath down the hall once a week. The reduction to necessity exhilarated her. She preferred the rigor and "what you see is what you get" tough realities to her parents' sumptuous Berlin house with its dark mahogany furniture, flower-patterned silk brocade draperies, Oriental rugs, and velvet upholstery fabrics with their silly gold fringes.

What Josef had left behind was nothing like Anni's circumstances of butlers and coach drivers. His mother—and, following her early death, his stepmother—did the cooking themselves; his father was a modest builder and overall craftsman who did plumbing, carpentry, electrical work, glass cutting, and house painting. But, like Anni's family's house, it was too cluttered. The materials were cheaper versions of the same stuff of which Anni's family had the expensive ones. The wallpaper was muddy and covered in flowers, the curtains grimly frilly and with scalloped edges.

Both the heiress from Berlin and the tradesman's son from the coal-mining city of Bottrop preferred visual and physical lightness. They exulted in it at the Bauhaus and maintained it for the rest of their lives. Linoleum, plywood, molded rubber, flat panels of glass: these were the Bauhaus's preferred ingredients, and the Alberses wanted nothing else.

The flawless, smooth steel-and-plastic casing of the iPhone provides the same satisfaction. The feat of accommodating a complex mechanism with remarkable capacity, all in a small and compact form, makes the iPhone a further leap. The respect for artificial materials, the wise dependence on chemical composites, is part of the continuum.

4.

Anni Albers used to say, "I don't understand all this emphasis on natural fibers. They can be as impractical and ugly as batik and macramé and all that other artsy stuff. I love 'drip-dry,'" which she pronounced "dah-*rhip*-dah-*rhye*."

Josef Albers marveled at the shiny tabletops, a thick wood encased in polyurethane, at his favorite restaurant, called The Plank House. It was hygienic and very easy to clean. At the same restaurant, he adored the salad bar. While it had red cabbage, which harked back to his childhood, its panoply of produce represented the triumph of modern transportation and the coordinated systems through which beans from Mexico and asparagus from Peru could be eaten in Connecticut. The clear plastic dome suspended above it enabled you to reach in and serve yourself whatever you wanted, seeing what you were about to take, but also prevented anyone's germs from getting directly onto the food; this was a marvel of modernism. Significantly, this was before the era when most people knew the consequences of these developments to climate change and their harm to the ecological balance of the earth.

When the Alberses took guests to the Plank House, they became practically rhapsodic. The innovations they admired at the Plank House, the impeccable synthetic surfaces and the generous range of experiences made possible, had the Bauhaus *spirit.* You could feel the mastery of process when your steak arrived cooked as specified. The new versions of original, well-known Bauhaus designs were only reproductions, after all. Anni and Josef had no nostalgia for life in Weimar and Dessau; the realities had been too fraught. What counted about the Bauhaus was that it celebrated *today:* the today of every era. To make the most of the present—and to engender excitement for the future—that was the true ideal of the experimentation, openness, and truthfulness that were the school at its best.

Anni Albers's bedroom in her and Josef's house at the end of their lives was so austere that it both jolted and transfixed you. The use of basic rolling window shades, of standard fabrication although meticulously cut to scale, was startling in the home of the foremost textile designer of the twentieth century. But while sophisticated people worldwide graced their elegant dwellings with Anni's drapery and upholstery materials, she and Josef had white plastic venetian blinds in their living room, Naugahyde on the sofas, and a gray industrial rug.

Anni's reverence for the word "plank," which she spoke about specifically, encapsulated what she and Josef prized. A plank of wood, she said, is simple in form and plain in surface. It is a by-product of good craftsmanship, a result of a skillful carpenter using a saw and plane and other tools to transform one of the most miraculous natural materials to accommodate human needs. The trees from which that wood is extracted are among the marvels of earthly existence. The need to have a solid place on which to put things is a basic human need. With a sound so delightful that one likes to repeat it, the monosyllabic noun "plank" had, for Anni, the enchantment of the material and processes, universal and timeless, it conjures.

The Alberses took the photographers Henri Cartier-Bresson, Lord Snowdon, and Arnold Newman to the Plank House. They also took the museum curator Henry Geldzahler and the newspaper and television journalists who came to interview them there. When they pulled up under the porte cochère at the entrance in their dark-green Mercedes 240 SL, the elderly German-accented couple, with Anni using a cane, and their exotic guests, comprised a strikingly different group from the other lunchtime clients. The local bankers, merchants, and shoppers at the nearby discount stores on the long shopping strip where the Plank House was located probably could not have imagined what the place was to the greatest Bauhaus artists still alive. In all likelihood, the other customers had no idea what the Bauhaus was. Anni and Josef, meanwhile, were not simply taking their guests for lunch. They were entering a temple of the new, of modernism for everyone, of friendliness and ease. They were celebrating the values they cherished.

Their guests were often open-mouthed at the ordinariness of the place chosen by the elderly artists, who were like gods to them. Lord Snowdon had imagined himself going somewhere that looked like one of the elegant Bauhaus masters' houses at the edge of a forest on the outskirts of Dessau. But to Anni and Josef, the Plank House embodied the same cultivation of the new, the harmony with the realities of place and time, the eschewal of quaintness or of architectural posturing. With its flavorful, tender steaks available at reasonable prices, the Plank House was also quintessentially Bauhaus in spirit with its capacity to enhance the everyday experience of people from different walks of life.

But in the charmingly named "Orange"—the Connecticut town where the Alberses lived—it was whiteness that mattered above all else. The Bauhaus headquarters, designed by Gropius in Dessau was, when it opened in 1925, white on the outside as well as within.

The masters' houses were the same, the exteriors and interiors consistent. Maybe the reason that the inside of the Alberses' house was such an eye-opener is because it was such a surprise after the humdrum exterior, like a treasure chest that on the outside is an ordinary cardboard shipping carton and that reveals its jewels only when you open the lid. Every wall was painted in the same white latex, which Josef tinted ever so slightly with a warm gray, in a way that no one else would perceive, but that he noticed in the way a great musician feels the most minuscule variation in tempo. While Anni's colorful textiles added their panache to lots of other people's homes, in her own house she had white plastic venetian blinds in the living room, dining area, and kitchen and white rolling paper shades in the bedrooms. You would never have known she had woven spectacular upholstery materials; the very ordinary living-room sofas, in a style we now call "mid-century modern," were covered in white Naugahyde, the kitchen chairs in white vinyl. There were no materials that could not be easily cleaned with a damp sponge.

The *Meisterhäuser* designed by Walter Gropius in a pine forest within walking distance of the Bauhaus put the Klees, the Kandinskys, the Alberses, and the Gropiuses themselves in larger spaces than they had ever inhabited before. And now they lived surrounded by the streamlined modernism the school advocated. But what was luxurious physically, and aesthetically pure, was not all it seemed to be. The Kandinskys used their old Russian samovar; the Klees' passion was their aquarium full of tropical fish. And the townspeople of Dessau were outraged that their government spent so much money on housing for outsiders, and so little on the local citizenry.

Whiteness of this level fills you with a sense of lightness and purity. It makes you feel clean. It opens possibilities. It is both a tabula rasa and a bright beginning, the way that a normal piece of typewriter paper is. And whether your iPhone is white, as the first ones were, or another color, the background screen is that same heartening and energizing white. It is not a regulation white—the warmth or coldness and precise tint are yours to determine—but what is vital is that it lacks added character. Essential for the legibility of whatever appears in front of it, that background, like the white of the Bauhaus, is also salubrious.

Anni was asked, over a quarter of a century after the fact, to write about her initial impressions of Walter Gropius after she arrived at the Bauhaus. She emphasized the whiteness she associated with him indelibly. It had both a visual impact and ramifications:

I came to the Bauhaus at its "period of the saints." Many around me, a lost and bewildered newcomer, were, oddly enough, in white—not a professional white or the white of summer—here it was the vestal white. But far from being awesome, the baggy white dresses and saggy white suits had rather a familiar homemade touch. Clearly this was a place of groping and fumbling, of experimenting and taking chances.

Outside was the world I came from, a tangle of hopelessness, of undirected energy, of cross-purposes. Inside, here, at the Bauhaus after some two years of its existence, was confusion, too, I thought, but certainly no hopelessness or aimlessness, rather exuberance with its own kind of confusion. But there seemed to be a gathering of efforts for some dim and distant purpose, a purpose I could not yet see and which, I feared might remain perhaps forever hidden from me.

Then Gropius spoke. It was a welcome to us, the new students. He spoke, I believe, of the ideas that brought the Bauhaus into being and of the work ahead. I do not recall anything of the actual phrasing or even of the thoughts expressed. What is still present in my mind is the experience of a gradual condensation, during that hour he spoke, of our hoping and musing into a focal point, into a meaning, into some distant, stable objective. It was an experience that meant purpose and direction from there on.

This was about twenty-six years ago.

Last year some young friends of mine told me of the opening speech Gropius gave at Harvard at the beginning of the new term. What made it significant to them was the experience of realizing sense and meaning in a world confused, now as then—the same experience of finding one's bearing.

5.

In essence whiteness is not so much a color as the visible absence of color; and at the same time the concrete of all colors; it is for these reasons that there is such a dumb blankness, full of meaning, in a wide landscape of shows—a colorless, all-color of atheism from which we shrink. . . . The mystical cosmetic which produces every one of her hues, the great principle of light, for ever remains white or colorless in itself. . . . And of all these things the Albino whale was the symbol.

—Herman Melville, "The Whiteness of the Whale,"
in *Moby-Dick; or the Whale* (1851)

The uplift and encouragement provided by whiteness in both Bauhaus and Apple surfaces derive in part from white being absent all hue. Hue is a prompt of specific emotions. Red, for example, can invoke the dangers of bleeding or the vagaries of romantic passion. Blue is many people's favorite color—the sky and sea in "perfect weather" (except for farmers in need of rain—but from there the mind takes off). Hue becomes personal.

Underlying whiteness, however, takes one into another sphere. Particularly when it is flat and unmodulated, white provides an ambient, inexplicable, nonassociative joy.

While Piet Mondrian never visited the Bauhaus itself, he was deified there, and his treatise on abstract painting was the first monograph by or about a single artist published as a Bauhaus book. Mondrian depended on whites, flat and unsullied, in the canvases he had begun to make in 1919. He had just moved to Paris following a five-year hiatus during wartime back in his native Holland. It was the same year the Bauhaus began, in a mood of postwar euphoria. Bauhaus artists who visited Mondrian were astonished by the pure white haven he had created for himself in his small studio apartment. He even had painted the wooden furniture white. He attached panels of primary colors on the walls in a way so that he could move them perpetually; the whiteness was essential as the backdrop for their activity.

Mondrian repeatedly used the term "the tragic" for what he wanted to avoid in his visual universe. It was not merely a matter of taste; it was an imperative. Whiteness was essential for setting the scene in which perfectly straight vertical and horizontal lines, and rectangles of primary color representing nothing but themselves, created the "neoplasticism" that was the direct opposite of the tragic. Other artists and architects who were not actually at the Bauhaus but who were developing in concord with it—Le Corbusier and Eileen Gray among them—also celebrated the optimism of whiteness and of visual priority on top of it. The iPhone and other Apple products depended on these ideals developed by sophisticated painters and architects.

Sure, you can proceed, with your iPhone, to any number of other things—to flowery dresses for sale or a panoply of available beach umbrellas or pornographic imagery—but that comes later. You start with the new aesthetic that reached an apogee in European design in the 1920s, with the advent of international modernism, and was cultivated at the Bauhaus as intensely as anywhere.

6.

The white backgrounds of the iPhone and all other Apple screens are matte and slightly tinted. They have a tiny amount of grayness to their cast.

The tone of the Apple screens has been formulated to accommodate the artificial illumination that comes from behind. It softens what would otherwise be glaring. The light that shines onto the screen is another matter. An issue yet to be successfully resolved in iPhone functioning, it can be problematic. Anyone who has ever tried to use an iPhone when bright sunshine falls on it, who cannot escape direct and intense daylight, has struggled with this. The light from within is modulated by the screen, but the glare from an exterior source can stymie the usage of the phone. The instrument becomes unreadable.

Recently, in January of 2019, there has been a major advance in iPhone screen technology. What Apple calls "liquid retina" has greatly increased the number of pixels per inch. The reading of the phone becomes sharper and clearer, and the new screen "supports a high color gamut." The language, which comes directly from Apple's advertising, describes a form of progress that would certainly have thrilled people at the Bauhaus, Josef Albers in particular. It increases the capacity to manage colors, enlarging the range of hues available. Especially in the late years of his life, when Josef was perpetually acquiring new tubes of paint made by the dozen or so manufacturers of top-quality oil pigments, and simultaneously exulting in the even larger range of colors allowed with printing ink, he would have thrilled to that expanded palette.

Josef had Albert Powell, his part-time gardener, meticulously coat the white surfaces of the backgrounds of his *Homage to the Square* series, applying between six and ten coats of liquid gesso, sanding each layer before putting on the next. Powell was the closest Josef came to having a studio assistant, differentiating him from today's successful painters with their large staffs executing much of their work. Those panels' background acquired a truer, brighter white than the iPhone's. So did the interior and exterior walls of the Bauhaus buildings and ceramics. But when the light on a surface, rather than coming through it, is projected onto it, the whiteness can afford such brightness. Marble, plaster, or painted wood, whether seen in sunlight or with the glow of candles or oil lamps or electric lighting, will never damage the eye.

White walls are hardly unique to modernity. They have been a norm in many cultures in many eras. But it is a mistake to think that the Greek Parthenon and its sculptures were the white that they are today.

The friezes and sculptures were painted in vibrant colors, some of the horses ruby red, with their riders in emerald-green boots. Goddesses with golden hair wore turquoise cloaks. The wooden columns of the Minoan temple at Knossos on Crete were also red. Almost all the color faded with time, however. The resultant whiteness suits the modern concept of purity in the classical era and the contemporary taste for absence of hue. In fact, in ancient Greece, as at the Bauhaus and with iPhones, the whiteness serves more than a single purpose; it offers a sense of a clean start, but it is also a necessary neutral background for the color action that follows.

Bauhaus whiteness, and the whiteness of computer screens, share their flatness. They differ from the whiteness of, say, the marvelous interiors of German rococo churches, or Michelangelo's marble *David*. In those cases, the surfaces were variegated. The church of Vierzehnheiligen is divinely white; whiteness is essential to its impact. But it is an example of plaster being metamorphosed to have the grace and airiness of billowing sails on a clipper ship, and the weightlessness and sheer joy of light meringue being whipped to frothy peaks.

This basilica, designed by the rococo architect Balthasar Neumann, was built in the remote town of Bad Staffelstein in Bavaria between 1743 and 1772. Josef Albers was ecstatic when he spoke about it. Everything about Vierzehnheiligen was the apogee of what Josef prized in art and architecture. It was gloriously white, a feat of engineering, light in mood in spite of the complexity essential to its realization, and devoid of any self-reverence on the part of the creator. It was made as an act of service—in this case to the worship of God and to the good deeds of Christianity, the sort of acts performed by the Vierzehnheiligen—the sainted band of fourteen Holy Helpers, whose innocent elation

Josef Albers loved good design and smart engineering in many styles. Modernism was not a requisite. The ability to use heavy materials to achieve billowing lightness, and the use of white to evoke cheer as well as purity, thrilled Josef in the rococo country church of Vierzehnheiligen. Its architect, Balthasar Neumann, joined the builders of Romanesque cloisters, and contemporaries like Le Corbusier and Alvar Aalto, as Josef's heroes.

the church makes manifest. (Toward the end of his life, Josef Albers attended mass almost every Sunday at the Holy Infant Church, a Catholic house of worship near his and Anni's home in Orange.) In rococo architecture you have the spirit of optimism; what we see is a source of incredible emotional uplift.

Vierzehnheiligen could not possibly be mistaken as having anything to do with Bauhaus *style*. But it fits in with what was essential at the Bauhaus. Its goal was to make life better through transformation that begins with what we see. What is solid can float. This happens when the walls and ceilings of a rococo church defy gravity, and when a small and dense handheld telephone/computer utilizes radio waves. Human life is ameliorated by the evocation of qualities beyond the self. Albers was so cranky about art that he considered too personal and at the same time aesthetically deficient that when discussing the rococo architect, he would add, "You must not confuse *Neu*mann with *New*man. Barnett Newman is not the same. The idea that those are 'the Stations of the Holy Cross!' Anni is right that his paintings look like bath towels. Too fuzzy, not enough there."

The denigration of certain art in which others of us see great merit was a consequence of an unwavering passion for what was deemed superior. It was not a matter of malice. The Bauhaus was not mainly a style, or a place, or a moment in history, or a set of facts; it was an attitude.

7.

Because white is achromatic, any and all other colors flourish on it. But not everyone sees it as neutral, and some go so far as to disdain it. People who dislike snow mind the white they associate with it, much as those who love snow relish it. When a bride wears white for her third wedding, with the children of her previous two attending, its association with virginity can raise hackles. Purity and chastity are, after all, what white signifies to most people. The pope wears it; so do Hindu Brahmans; so do Muslim marabouts.

It can also be the most reverent color for according significance to death. Serge Diaghilev, the great ballet impresario, requested three of his closest female friends, when they visited him on his deathbed in Venice, to wear white at his funeral. They did as instructed, with style. Coco Chanel, Misia Sert, and the Baroness d'Erlanger all looked ravishing in their white dresses as they followed Diaghilev's black catafalque, transported in a black gondola (the Venetian form of hearse) to his funeral at the church of San Giorgio dei Greci. If he wanted it to seem that he was already being transported to heaven, as mass was sung for him in that icon-fitted church—of greater significance to El Greco than was the Italian Renaissance art he had gone to Venice to study—he succeeded.

White is also—and this is a central element of its pertinence both to a lot of products of the Bauhaus and to the iPhone—sanitary. Because white surfaces show dirt, they are easy to clean, and then to recognize as pristine. Cleanliness is something most people find reassuring (although that, too, is a personal taste).

The white that is the background of computer screens, iPhones among them, functions like the white of a clean piece of typing paper, or of the "supports"—to use the correct art-historical term—of the paintings Klee and Kandinsky and Schlemmer made at the Bauhaus. Whether they were working on paper or canvas, white was necessary not just as the recipient of and most useful background for color, but also because of the way it is a start. And most important, it is otherworldly.

8.

White, after all, is the color, or noncolor, of the universe. The white of the Bauhaus and of iPhones is not as lofty as these sacred whites, but it is, as they are, removed from everyday clutter, whether the clutter is physical or mental. And it wakes one up. Jobs opted for it for the same reasons that Gropius and his colleagues did. This common element of the school that codified modernism and the screen of the object now used worldwide is unblemished, inspiring, devoid of association, modern, and intrinsically uplifting. It declares itself the basis for whatever action comes next, a clean slate, with the problems of the past irrelevant. And it brightens your mood. Why white increases serotonin levels probably eludes neurobiological explanation, but that lift is real.

The white screen allows maximum legibility while energizing the user.

The world most of the Bauhauslers came from was like that of the two Alberses, however different the variations. Europe at the end of the nineteenth century and the start of the twentieth was still coated in the sooty black of industry; everyday objects tended to be heavy like the crimson velvets on Biedermeier chairs, or saccharine like the pastel tones of cuckoo clocks. The white the Bauhauslers coated things in, and wore, in Weimar and then in Dessau, had a wonderful absence of history. It put one in the present and primed one for the future. Steve Jobs's world, before he began to make things white, was also laden with associations, murky history, and confusion.

The whiteness fundamental to all Apple products similarly clears the mind while launching fresh ideas. The surface one sees and touches on a wireless Apple computer mouse, a uniform tone like that of a white eggshell, the surface nicely hard but neither brittle nor fragile, the form accommodating to the human hand, is unencumbered and inviting. The back of the mouse, which one does not see, is an anodized aluminum with a couple of flat black rubber bars that create the base. You may consider it a stretch to compare those bars to the window trim in Dessau, and the matte gray of the underside of the mouse to the railing of the balconies there, but the staccato beat and solidity are similar.

That little device that controls movement on the computer screen and the large building that contained all of the Bauhaus workshops have in common that they are an invitation to take action. The future is yours; this is an opportunity to go some-where. You are the one in charge; the building, like the small tool, is a vehicle for your convenience. That fundamental applies to the iPhone as well.

Part VI

1.

The workings of the iPhone echo activity within the Bauhaus in their sheer rapidity. What happens happens lickety-split—but only as a consequence of a lot of reflection and meditation first, and of what we now call "mindfulness." The term has become irritating because of the inevitable lugubriousness with which it is uttered, and because of the implicit suggestion that no one before, in all of history, has actually concentrated on what he was doing and expanded his awareness through contemplation. Still, focus and careful planning precede quick construction and then activation. Then the building goes up, or the weaving gets created on the loom; the intelligent preparation allows action and spontaneity to follow. The development of the iPhone was long in coming. Every aspect took years. The internal mechanism was revised time and again. Its housing went through an incalculable number of propositions and rejections. Once the basic shell was selected and approved, modifications and refinements ensued. The packaging and the marketing strategy followed, none of it fast or simple. But once the end result was launched, the iPhone was out there to be sold quickly and in volume. Then, as soon as it is bought, the user charges it, turns it on, and knows that every pressing-down of a fingertip will bring instant results.

Speed is part of the essence of modernism. Jet airplanes fly us across vast oceans almost as fast as the earth rotates. Television brings, in an instant, events that occur in China to people watching them in France or Brazil. We no longer expect to wait for anything. Our predecessors, in earlier centuries, lived at a different pace. The potato that gets cooked completely in ten minutes in a microwave oven took three hours when the only way to make it edible was in a wood fire that first had to turn to smoldering coals.

The experimental Haus am Horn that was part of the 1923 Weimar Bauhaus exhibition went up in four months. Chairs in Dessau progressed from design to completion in weeks. Stainless-steel door handles required elaborate conceptualization, but the journey from fabrication to installation to use occurred in days. After all those years of development, the iPhone gives us, right away, the family photo we want to show a friend or the YouTube video we insist must be viewed immediately or the boarding pass that gets us onto the plane.

The new words that developed with the first computers and all of their variants have now become part and parcel of modern life, with the iPhone having made their use even more widespread. But, like Bauhaus design, before they had been invented by a few intrepid word-jugglers, the larger populace could not have imagined them. The vocabulary that includes "screen savers" and a "mouse" that is not a rodent is revolutionary. So is the imperative to "log in" by using your thumbprint or your private password. The capacity to make our iPhone our own is essential to the independence all human beings want, even as we insist on interdependence through these devices that make us instantly accessible to others.

Not so long ago, the Bauhaus baby cradle, made by two rectangular wooden slats meeting in a V formation and suspended between two flattened cylinders, was equally radical. The iPhone initially looks just as simple. Then, with human action, everything changes. The communication facilitated by that simple object in our hand becomes as complex and unpredictable as life within that cradle once it holds a baby. And that richness—of life itself—facilitated through simple mechanisms, merits celebration. To have access to the panoply of human experiences while appreciating their fundamentals was the Bauhaus dream.

Through technology and good design, the wonders of the universe—whether as general and timeless as centrifugal force and the miracle of the sun or as specific as the need of the moment—are made accessible.

What is wanted from the tools is the user's choice. Bauhaus kitchens make it easier to prepare meals quickly and simply, throwing together ingredients and cooking them almost instantly, or to create a croque-en-bouche with elaborate pastry and filling and frosting all made from scratch. Your iPhone allows you to have urgent, vital communication or to meander via "social media" (horrid term) and reestablish a connection with people you have not seen in fifty years. The point, as with Bauhaus design, is not to impose on you, but to enable you to fulfill your needs and wishes.

2.

The facing of the iPhone—the flat wraparound screen of remarkably thin and durable glass—is one of the smartest choices behind the success of the device. The absence of a frame confining it echoes the lack of paneling or molding on the walls of Bauhaus interiors. Flatness presides uninterrupted. And the capacity to act is strengthened by the lack of distracting elements.

Like a lot that Steve Jobs insisted on, the screen rendered useless the intense labor that various people had devoted to the project design. Jonathan Ive, who had by then become Apple's chief designer, had made an initial prototype of the phone with a glass screen in an aluminum case. Jobs had encouraged these efforts; then, after the Apple team had been working on Ive's design for nine months, he changed his mind. Jobs decided the problem was more about the case than about the display. Again, like the Bauhauslers, he bowed to no one where his standards and vision were concerned, and was willing to revise endlessly until he attained perfection—rather than settle for second-best.

Ive accepted Jobs's challenge to redesign the phone so that the Gorilla Glass display went right to the edges. Like Jobs, and like Gropius, Ive was one of those rare people for whom the search for the new and the better came as naturally as breathing. These visionaries were driven toward the material realization of the lightness and freshness they sought in their own lives, and they would do what it took to get there. The rock-hard, paper-thin, flawless glass on which Jobs insisted and which Ive refined is so light, physically and visually, that it seems to float. Its wonderfully delicate stainless-steel frame holds it not like a tightened vise, clutched in the grip of despair, but, rather, like a feather barely touching the surface over which it floats. It is the gentle touch insisted on by sports trainers when they say to hold the framework of your gym machine only so you are situated correctly, not to grab it as if for dear life, which would defeat the point of the exercise you are doing for another part of your body; it is for position only.

Gorilla Glass was a new product of Corning Glass. In the same time period, that company in upstate New York commissioned Josef Albers to make murals for the lobby of their sleek and elegant new headquarters in Midtown Manhattan. Composed of lines fabricated in thin, round bars of machine-perfect stainless steel, held by invisible supports in front of a marble wall, these geometric drawings initially appear simple but in fact never stop moving and visually transforming themselves. Our eyes perceive them first as open boxes moving in one direction, and then as the exteriors of those boxes of which we just thought we were seeing the insides. At one moment, we are on the verge of entering rectilinear caverns; at another, we are blocked by solid crates. That this occurs on flat walls, simply because of the careful articulation of a few straight lines, is an example of Josef's beloved "minimal means for maximum effect." With limited elements, we are beckoned into an endless course of illogical and impossible experiences. This was the artist's fondest intention—which he also realized with his seemingly simple paintings in which flat, solid colors interact to create a range of shades and echoes, physically impossible but optically vivid.

Reproducing a Bauhaus master's carefully conceived abstract/figurative drawings or producing, for the iPhone, the new Gorilla Glass—that name suggesting one of the strongest and most dangerous of wild animals—Corning used the capacity developed in its laboratories to make exceptional materials and fabricate them masterfully. The collaboration with modern industry—its technological advances in this case applied not to the tools of war, but to the enhancement of life in peacetime—facilitated wonderful possibilities in the hands of good designers.

Corning had, in 1960, produced the reinforced glass they called Chemcor. Its salient trait was that it was practically impossible to break. To make it, glass was dipped into a hot solution rich in potassium salt. This caused smaller sodium atoms in the glass to be replaced by larger potassium atoms.

The tight compression of these large potassium atoms against one another, which occurs once the glass is cool, makes it so tough that it can stand up to the impact of a physical shock of a hundred thousand pounds per square inch. It was useful for prison windows as well as phone booths. But when it was being tried for car windshields, the crash tests showed that even though the windshield survived, the human skulls hitting them were more likely to shatter than against other materials. Besides, it was prohibitively expensive. About a decade after its invention, Chemcor was discontinued.

In 2006, when Steve Jobs approached Corning about redoing this glass with such remarkable powers of endurance, he passed on to the people at the glass company the specs that Jony Ive had provided. The glass, he said, could be no more than 1.3 millimeters thick. Corning replied that the task was impossible: they refused Jobs's request to put a material they deemed commercially worthless into mass production. He did not accept their answer. His deadline of six weeks and the quantity he wanted were both laughable.

Jobs's persuasiveness in getting done what others said couldn't be done was another similarity with the leading lights of the Bauhaus. By May 2007, in the factory where they formerly had been making LCD displays, Corning was producing thousands of yards of what they now had renamed Gorilla Glass to be the face of the new iPhone.

The back of the phone was aluminum. A stainless-steel bezel held it and the glass sandwiched against one another. A thin rubber gasket inside the bezel and the glass screen served as a sort of shock absorber. The thickness of the glass adhered to Ive's precise measurement as specified by Jobs.

In 1934, when Josef Albers designed, at the request of the

architect Philip Johnson, the cover for the catalogue of the Museum of Modern Art's *Machine Art* exhibition, brandishing a dramatic photo of an industrial ball bearing, Josef rejected the initial proofs from the printer, demanding that his name as designer be removed if its flaw was not corrected. Johnson wrote back contritely. He assured Josef that the margin between the edge of the paper and the powerful image of that glistening, perfectly functioning machine part would be reduced to a precise three millimeters in accord with Josef's original design. Johnson understood the gravity of the mistake and the importance of measurements. These were among the Bauhaus standards that the ultra-fine but sufficiently tough Gorilla Glass maintained.

On the day when the iPhone was made public with great ceremony and copious publicity, Jobs took the time to write the man with whom he had worked at Corning to say, "We couldn't have done it without you." The glass made it possible for the overall iPhone not just to be incredibly thin, which is why it fits so easily into a pocket, but to open onto its unblemished rectangular field of clear luminous whiteness. The frame softens, and nearly obliterates, the edges and corners of the phone, making it more agreeable to the human touch. Were its edges sharp, the user would be less comfortable. Bauhaus objects were designed with the same intended ease.

Of course most iPhone users are reading some of these details skeptically. Almost everyone who has had an iPhone long enough has cracked the screen. And certainly the jolt that caused it to break was not even a fraction of the hundred thousand pounds per square inch that Gorilla Glass is supposed to be able to withstand. Still, these objects can survive a lot of abuse.

3.

The house and the aesthetic surroundings in which Steve Jobs grew up lacked Bauhaus perfection, but, like the architecture of Gropius, it was lean and functional. The suburban home where Paul and Clara Jobs raised their two children (they adopted a baby daughter, Patti, two years after Steve) was well conceived and light in feeling. Joseph L. Eichler, born in the Bronx in 1900, had developed the formula of which their single-family dwelling was a clone. Its style belonged only to the present, and it was geared for productive living.

Eichler was a prosperous businessman who had moved to California. His wealth from his family's dairy business enabled him to rent a luxurious house by Frank Lloyd Wright in Hillsborough. Settled there with his young family, he reckoned that he could make a second fortune if he developed a home where people without the financial means to own a Frank Lloyd Wright original could enjoy the same aesthetics without the frills. Eichler figured that the style Wright had perfected in his California houses could be echoed and produced in quantity. There would be no individual architect to drive up the costs; builders could mass-produce similar dwellings that would suit a range of clients.

Eichler's prototype did not depend only on the American Wright. Mies van der Rohe's domestic style—as it was manifest in his exquisite designs for his unrealized brick-and-concrete country houses, as well as in the Villa Tugendhat and the Barcelona Pavilion, all from the Bauhaus era—was essential to Eichler's vision. Most of Eichler's houses had flat roofs, exterior walls that were floor-to-ceiling glass, and a prevailing whiteness inside, with visible rafters painted to seem feathery-light and luminous. Eichler used California redwood and local bricks.

The houses varied in their details but shared their complete lack of ornament and their overall openness.

Eichler's company built over eleven thousand tract houses. But they directly influenced, in California alone, over a hundred thousand single-family dwellings that popped up in the 1950s in a style dubbed "Likeler." The Jobses' house was one of them Steve Jobs spoke reverently of these pleasant, low-cost, functionalist creations to Walter Isaacson when they walked around the neighborhood where he grew up: "Eichler did a great thing. His houses were smart and cheap and good. They brought clean design and simple taste to lower-income people. They had awesome little features, like radiant heating in the floors. . . . I love it when you can bring really great design and simple capability to something that doesn't cost much."

His childhood home was where Jobs's passion for innovative, straightforward, and affordable design began. "It was the original vision for Apple," he told Isaacson. "That's what we tried to do with the first Mac. That's what we did with the iPod. . . . It takes a lot of hard work to make something simple, to truly understand the underlying challenges and come up with elegant solutions. . . . I have always found Buddhism—Japanese Zen Buddhism in particular— to be aesthetically sublime. The most sublime thing I've ever seen are the gardens around Kyoto."

Jobs had initially studied Buddhism in India when he was nineteen. He began to meditate, and was soon drawn to the Japanese version of the practice. It helped him calm his mind and increase his receptivity to subtleties and his faith in intuition. His attraction to Zen only intensified when he became enchanted by the rock and moss gardens and other miracles of refined simplicity and poetic grace that reached an apogee in ancient Kyoto. It would be hard not to succumb to the gloriousness of their visual rhymes and rhythms, achieved with the restraint of a Bach par-

This classic Eichler house shows the prototype of the modernist, one-story houses that proliferated all over California in the 1950s. Joseph Eichler, the developer who built and popularized them, had lived in a house by Frank Lloyd Wright, and adapted its style to single-family dwellings that were clean-lined and relatively inexpensive. They brazenly put the garage in front as if to emphasize that the way of life they accommodated was completely dependent on the automobile and its capacity for people to go substantial distances quickly. The house in which Steve Jobs grew up was a spin-off of an Eichler, one of many thousands that sprung up in the era, and he relished its impact on his taste for lively, minimalist, straightforward design.

tita. Zen refinement and rigor became vital to Jobs's sense of design. The impact of Joseph Eichler—whose homes now fall in the category of "mid-century modernism," with the fanciest ones in places like Palm Springs, commanding hefty prices—was equally profound. The house in which Jobs grew up transmitted seminal values to him. He absorbed an approach to life that was truly Bauhaus, if not so rigorously Zen. Subsequently, its attitude penetrated all Apple designs, with the iPhone the quintessence.

Eichler was not simply a property developer. He was a genuine idealist. He wanted middle-class society to flourish, with a new aesthetic refinement penetrating American life. His goal was to construct planned neighborhoods where parks and community centers served one and all.

This followed the example recently realized by Le Corbusier's triumphant apartment building in Marseille. Built for people of more limited financial means than Eichler's client base, Corbu's "Unité d'habitation" had affordable units that, in one direction, gave a view of the mountains, in the other of the sea. Every inhabitant had access to a municipal garden, a nursery school, food shops, and guest rooms for visitors. Eichler was remarkable for his insistent disregard of religion or race; he had conspicuously resigned from the National Association of Home Builders when they would not adopt a policy forbidding discrimination. His stance opposed the practices of another major developer of middle-class housing, Fred Trump, father of a future president of the United States. Trump's firm, on the East Coast, used a small *c* to indicate "colored." It was penciled onto the application of any black family seeking residency in one of his buildings. Their applications were then rejected with some other reason given, the *c* erased.

The iPhone was, of course, made for one and all. This was motivated above all by financial goals—the more sold, the better—but Jobs's politics were similar to Eichler's. With his own background both Muslim and Christian, his approach to society at large was completely egalitarian.

The Bauhaus, institutionally, was trickier on these issues. There were no blacks at the school. Jews were referred to as "non-Germans" in the censuses required by the government; they were never to exceed 10 percent of the total number of faculty and students.

The excuse, often given, of "Oh, you have to consider the place and the era" is no justification for this horrific rule. People breaking through tradition in the field of design should have done the same with religious and racial prejudice. Josef Albers was exemplary. He knew that his father disapproved of his marrying a Jewish woman. His conclusion was not to invite his father or any other of his blood relatives to his wedding in Berlin. Anni's parents organized the marriage ceremony in a Catholic church because Josef, who was devout, wanted it there. The setting would have suited Lorenz Albers, but Josef would not tolerate the presence of the father and the one of his two sisters who disdained Anni's Jewish background. His only regret was that this prohibited inviting his other, unbigoted, sister to attend.

But the consistency of Josef's progressive stance on art and design with his attitudes toward race was not universal. Wassily Kandinsky's anti-Semitism and his doltishness about it was so deeply hurtful to Arnold Schoenberg that Schoenberg turned down an invitation to teach at the Bauhaus. Marcel Breuer concealed his Jewishness totally and had himself made officially "Aryan." Mies van der Rohe blithely mocked "spoiled Jewish girls from Frankfort" (in Anni Albers's presence). And the Bauhaus administration adhered to the 10 percent Jewish quota rigorously and without challenge.

Still, Bauhaus designs—like Eichler's and Apple's—were egalitarian by plan. They differ from automobiles and other material objects where the rich have one thing and the poor another, or where what you own signifies your wealth and confirms your social status. A student with financial struggles and a billionaire might often have identical Bauhaus furniture, albeit with the student's version being an inexpensive copy, or equivalents of iPhones, even if they were variations significantly less expensive than the authentic ones. Of course not everyone in the world can afford the sanctioned Bauhaus furniture or actual iPhones but the goal underlying the creation of these objects was to unify people by giving them access to designs that could be adopted for the use of almost everyone.

Part VII

1.

For all that gets told and retold about the Bauhaus, hardly anyone today is aware of how rapidly it became revered internationally soon after its founding. Word spread about the school throughout Europe, North and South America, and Asia. Its new approach to design, the quality of its teaching, and the art and objects being made there captivated people in every echelon of society during this period of looking forward that had begun following the peacetime accords of 1918 that ended the world war. In science, music, medicine, poetry, psychology, and other realms, human civilization was in a growth spurt, and what was happening in Weimar was revered.

It would take longer for the Bauhaus to become known, or to have an impact on the look of things, in Africa. Even though it was not so far from Weimar, the large continent to the south of the Mediterranean was still kept apart from most of the rest of civilization, except as a provider of raw materials and cheap labor. But elsewhere the Bauhaus was hot.

The first substantial presentation of work from the Bauhaus was in Calcutta, in December of 1922. It was part of the Fourteenth Annual Exhibition of the Indian Society of Oriental Art, a show that exceeded the purview of its title. Inaugurated by the governor with a Hindu ceremony, this exhibition, which on one side showed Bengal paintings of Shiva meditating on the Ganges and other traditional images of gods and goddesses, presented in the other half of the exhibition space the latest output from the three-year-old art school in Weimar. It attracted a large audience. The enthusiastic visitors ranged from dignitaries and socialites and business leaders to poets and artists. The greatest draws were information on the school's philosophy and recent paintings by Kandinsky that were exuberant manifestations of the Bauhaus. Not everyone approved—some viewers considered the school's goals too Bolshevik, and were shocked by the purely abstract art—but no one ignored it.

The next Bauhaus show was the vast comprehensive exhibition the following year at the school itself. It attracted some of the most creative people of the era, among them Igor Stravinsky. And Bayer's catalogue further expanded the international audience.

By the end of the 1920s, a number of sophisticated Americans had visited the Dessau Bauhaus—including the architect Philip Johnson; Alfred Barr, the future director of the Museum of Modern Art; and Edward Warburg, a young patron and collector. They brought back to the United States glowing reports of the breakthrough creativity they witnessed. In 1931, the Harvard Society for Contemporary Art, cofounded in 1928 by Warburg, Lincoln Kirstein, and John Walker, all Harvard undergraduates, in two rooms they rented above the Harvard Coop, put on the first Bauhaus show in America. It was a superb small exhibition with objects fresh from the school's workshops.

After the school closed in 1933, Bauhauslers fled to the United States. In 1933, Josef Albers had begun to teach at the newly formed Black Mountain College in North Carolina. He made certain principles of Bauhaus experimentation essential to the education he advocated all over the country. László Moholy-Nagy, another Bauhaus professor, had, in 1937, become director of what was nicknamed "the New Bauhaus"—the Association of Arts and Industries—in Chicago. Walter Gropius, who had been the first person invited to assume the position, had recommended Moholy-Nagy for Chicago because he had taken a teaching position at the new Harvard Graduate School of Design and, the following year, became chair of its Department of Architecture.

Exhibitions would follow at the Museum of Modern Art and other important institutions. Most art movements emerge and die; the Bauhaus entered the lifeblood of American culture. And many of its best-known designs proliferated.

2.

The transmission of Bauhaus values to the construction and design of the iPhone was substantially through Jonathan Ive. The chief designer of the iPhone had an education and training that were Bauhaus through and through.

Like Jil Sander, Ive shortened his traditional first name to a rare and strikingly modern one. Jony Ive, who attended the same school that David Beckham would go to eight years later in Chingford, an affluent suburb in northeast London at the edge of rural Essex, was not mainstream. For one thing, he had dyslexia, which made school unusually difficult for him. Like Steve Jobs. who was also dyslexic, his difficulties in developing certain rudimentary skills contributed to his precociousness in other realms. His parents' professions were a boost. His mother, Pamela Mary

Ive, was a psychotherapist. Jony's father, Michael John Ive, in addition to being a silversmith, taught craft and design in Jony's school. They both helped Jony learn handicraft skills.

Then Mike Ive—who also preferred his nickname, albeit a familiar one—was appointed Her Majesty's Inspector for the Education Ministry, with the responsibility of overseeing the teaching of design and technology in his district. Once he attained that post, he elevated the status of design technology throughout Great Britain by making it part of the core curriculum in state schools. Mike Ive helped enact the legislation that enabled all students aged five to sixteen to take classes in a subject that had previously been considered too esoteric. An aficionado of the Bauhaus, he made Bauhaus teaching methods, specifically, central to Jony's formation.

At the same time that Mike Ive elevated design technology and made it the equal of science, math, history, and the other core school subjects, he did the same as a parent. Ralph Tabberer—a colleague of his who would become Tony Blair's director general of schools—describes Mike Ive "constantly talking to Jonathan about design. If they were walking down the street together, Mike might point out different types of street lights in various locations and ask Jonathan why he thought they were different: how the light would fall and what weather conditions might affect the choice of their designs."

At school, the dyslexic Jony was a good draftsman, with a remarkable capacity for verisimilitude. The head teacher papered the walls of his office with Jony's drawings and watercolors and encouraged him to cultivate his strengths, not bemoan his limitations. Jony's father had him consider how he might improve the objects of everyday life. As a teenager, preparing for the A-level exams required to apply for university, Jony designed new models of mobile telephones.

Jony was outside the norm on many fronts. A drummer in a rock band, he wore his hair in "a shoulder-length mullet with a fringe that was back-combed to stick straight up," so that people said he "looked like a hairbrush." On the more diligent side, he made a model of a portable overhead projector that his teachers entered in a national competition, with Terence Conran the judge. Advanced techniques for projecting images had been a passion at the Bauhaus as well; the auditorium for which Anni Albers made her diploma material had featured a breakthrough use of three such devices made by Zeiss. The winner would get the "Young Engineer of the Year Award."

Jony's projector made it to the second round. Hoping to go further in the competition, he took the model apart to perfect it. When he reinserted the lens, he put it in backwards. The projected image looked like pure mush, and Jony fell out of the running.

Still, Mike Ive asked the managing director of the Roberts Weaver Group, a large industrial design firm, if RWG would sponsor Jony at Newcastle Polytechnic. Jony's A-level results were so high that he could have attended Oxford or Cambridge, but Newcastle Polytechnic was supreme in industrial design.

Jony was exposed in even greater detail to Bauhaus methodology at Newcastle. In his first year there, he entered a contest sponsored by Sony, for which he made a fantastical landline telephone. Jony called it "the Orator." Made entirely of a white plastic tube one inch in diameter, it resembled a question mark, with the top curve holding the earpiece. It got him a five-hundred-pound travel award. From there, Jony went on to design hearing aids and then a flat-screen white plastic ATM, taking prizes with almost everything. The streamlined cash dispenser won him fifteen hundred pounds from Pitney Bowes.

3.

The link from Weimar and Dessau to Cupertino was direct. The design writer and professor Penny Sparke, vice chancellor at Kingston University, explained to Jony Ive's incisive biographer, Leander Kahney, the direct connection. "The German Bauhaus of the 1920s was picked up by British design education in the 1950s. For example, they had what was called a foundation year in Bauhaus, and British design also had a foundation year. The idea of the foundation year was that students started from scratch; they did not build on the past but started on an empty page." Jony Ive was educated accordingly.

Josef Albers taught that foundation course at the Bauhaus. He would continue promulgating its principles in the United States through his teaching at Black Mountain College and the Yale University School of Design. Albers also excited audiences of artists, architects, and graphic and industrial designers with the possibilities of simple but imaginative construction at institutions throughout the States as well as in South America and at the influential Hochschule für Gestaltung in Ulm, Germany.

In one core exercise, students would take single, flat sheets of paper and fold them in complex ways. Manipulating that one thin plane of material, they achieved surprising results. Jony Ive developed his feeling for form and design with this exercise.

Students often started with the accordion fold. They first positioned the plain white sheet of paper horizontally. Next they made, at small increments, sharp straight creases, between top and bottom, first in one direction, then in the opposite direction. Rather than reinforce those creases, the students would go over them a second time, reversing the direction of each. This made the creases function as hinges, able to open and close, rather than permanent folds.

Where the students went next was up to them. They might compress the sheet with its accordion folds so that it was like a closed fan. They might subsequently fold that closed fan in half in a direction roughly perpendicular to the many straight folds so nicely aligned. They would discover that this crease to halve the overall construction was more effective at an angle than straight across. If they then opened up the resultant object, it had the effect of a bellows. The movement of air added complexity.

Opening and flattening the paper provided endless possibilities. Students might fold the sheet top to bottom, then side to side, and reopen it, either totally or in part. Or they might fold the sheet at an angle and work with triangular forms. Albers did not teach them to follow instructions as one would with an origami "how to" list of commands; rather, he would encourage them to go their own way. What was fundamental was making the most of minimal materials. Good design required concentration and focus, and the will to do the unprecedented.

With paper folding, Albers encouraged a keen awareness of materials, of their pliability and their textures. Nothing was wasted; everything was stretched to capacity. Light, and shadow effects, needed to be considered from the start.

The insistence on a reduced vocabulary of elements and minimal but effective constructive methods would become central to both Steve Jobs's and Jony Ive's approach. They would adhere to the tenets of the Bauhaus foundation course.

Self-expression was to be avoided. The essence of the Bauhaus design teaching that helped form Jony Ive was that what mattered in drawing technique and color exploration were the properties of the components and methods, not a personal narrative.

Hannes Beckmann, who studied at the Bauhaus and later taught art in America, recalled:

The first day of the Preliminary Course, Josef Albers entered the room, carrying with him a bunch of newspapers . . . [and] then addressed us . . . "Ladies and gentlemen, we are poor, not rich. We can't afford to waste materials or time. . . . All art starts with a material, and therefore we have first to investigate what our material can do. So, at the beginning we will experiment without aiming at making a product. At the moment we prefer cleverness to beauty. . . . Our studies should lead to constructive thinking. . . . I want you now to take the newspapers . . . and try to make something out of them that is more than you have now. I want you to respect the material and use it in a way that makes sense—preserve its inherent characteristics. If you can do without tools like knives and scissors, and without glue, [all] the better."

Jony Ive's first design to garner approval could have come out of that class. As a student at Newcastle Polytechnic, he was given the task of developing prototypes for wallets for a Japanese pen manufacturer. Ive made his out of white paper, which he folded and refolded and cut so that it opened and shut somewhat like an accordion file.

He would use miraculously thin materials for the iPhone—front, back, and sides—and would bend aluminum in keeping with his early teaching. To realize infinite possibilities from a minimal vocabulary of structural elements was a stupendous skill.

4.

The Bauhaus foundation course was also taught by Johannes Itten and László Moholy-Nagy. They and Albers were very dis-

The preliminary course Josef taught at the Dessau Bauhaus, requisite for all students, emphasized paper folding as a means of understanding the imaginative manipulation of pliable material to create forms in space. This image by the photographer Umbo provides a glimpse of Bauhaus life, with the few women dressed like prim schoolgirls and most of the men in suits and ties, except for one raffish character in three-quarter-length golf pants. Jony Ive's father would teach a similar approach; although Jony, a drummer in a rock band, looked nothing like a Bauhausler, his first successful design could have come straight out of this course.

similar as artists, but as teachers the three of them emphasized not just the need for students to understand and respect the true nature of a vast range of materials, but to take pleasure in them. Contrasting surfaces should be savored as sources of tactile and visual delight. Meanwhile, the different capacities of plastic, wood, hemp, silk, jute, velvet—deliberately combined or opposed—opened an array of means of constructing things. Good designers understand how best to achieve elasticity, solidity, tensile lightness, or compact mass. They are open to synthetics as well as raw substances. Bauhaus education went against the tradition of the academies with their insistence on singular methods. No one was told that this color goes only with that one, or that a human figure must be drawn as if for an anatomy book. What was encouraged was the development of practical knowledge and the will for risky experimentation.

The technical and programming aspects of the iPhone were the domain of other people at Apple. Jony Ive's task was to make it a seductive, functional object. To do so, he would experiment with known materials and develop new ones. The feeling for *"matière,"* which Ive came to initially through his father's Bauhaus teaching methodology, was vital. Ive also adhered to another taste of the Alberses', which was the openness to surprises. Playfulness was taken seriously. Josef and Anni liked materials to have multiple functions—to be entertaining as well as practical—and to jolt the user with unexpected twists.

A psychoanalyst Anni knew—a traditional Freudian who was a Connecticut neighbor—asked her if she could imagine a textile he could work on during treatment. Anni relished the challenge. The doctor needed to work in a way that his patients could not perceive. The patients were lying on their backs, facing away, at a right angle to the doctor, who was seated upright. The doctor's task had to be diverting, even calming and pleasant, but not distracting. She created a colorful gouache of a playful pattern of irregular forms a bit like flagstones. The doctor could work on each of the solid-color oblong forms singly, knowing they would be arranged in a playful and rhythmic mosaic.

It was important to consider new approaches and solutions to human needs, ideally ones that added levity to the daily routine. Josef Albers made a simple teacup out of ebony, glass,

Josef Albers made this simple tea glass and saucer in three different Bauhaus workshops: glassmaking, carpentry, and metalwork. It was exceptionally user-friendly, with the vertical ebony handle positioned for the person serving tea, and the horizontal one for the recipient. The refined use of various materials to facilitate easy use as well as pleasant tactile feelings was essential to the design of the iPhone.

and stainless steel that required his making the refined elements in three different workshops. The thin, flat steel band around the simple glass, a squat cylinder with lips that made for easy sipping, had two handles at opposite sides. The handles were identical flat ebony disks, smooth and pleasant to touch, and ideally proportioned for human fingers. But one was positioned horizontally, one vertically. The upright handle suited the ideal grip for the person handing over the glass of tea. Albers had studied the anatomy of the human hand and its relation to arm movement. When extending the forearm in the act of offering tea, it was better to press thumb and fingers against one another on the two sides of the vertical disk. Pulling the cup toward oneself, it was better to grasp the disk that was cantilevered from the steel band. The utter originality and accommodation to the purposes of design in both Anni's rug and Josef's teacup were the essence of Bauhaus ingenuity and generous spirit.

Part of Jony Ive's mythology is that he next designed a remarkable pen for Zebra—the Japanese company for whom he had made the wallet that folded so cleverly. The unusual pen went into production with the name TX2. It functioned impeccably as a writing implement, with rubber extrusions on the shaft that gave the user a better grip. Ive is credited with cleverly accommodating what he called "the fiddle factor" with a small clip holding the ball on the top. The ball rotated. Ive is said to have observed the way that people holding pens often tend to fidget with them, and so to have included this element only for the user's diversion and release of nervous energy. The playful ball that served no other purpose would not break off the way a pen clip would. The emphasis on improved utilitarianism in the protrusions that steadied the user's grip, and on the enjoyable in that unexpected ball, reflects the awareness of human needs shared by the avatars of the Bauhaus and the future designer of the iPhone.

The pen is great design, and adds luster to Jony Ive's history as a designer. But when the William Weaver Company was approached about reproducing one of their licensed photos of the pen for this book, Barrie Weaver, the head of the company in Bath, England, vehemently refused to grant the rights. Emphasizing that the Weaver Company sponsored Ive through university, and had him as part of their design team for almost two years, Weaver wrote that journalists have failed completely to grasp the way that this, and other designs, evolved. (For this reason, Barry Weaver also refused to grant the rights to reproduce their photo of Ive's Orator.) Weaver made clear his resentment for what he considers Ive's grandstanding about the design of the pen, and for the attribution of this original and clever object to the iPhone designer alone in the many sources, in print and online, where the pen appears. Weaver was enraged that Ive has become the only person given credit for a product that was the result of a collaborative effort, since the truth is that Ive was just one of many people in the team that developed the pen. Besides, the execution of the pen depended on the work of first-rate engineers in Japan.

Weaver could not be persuaded that this book would do better than all the other sources that credited Jony Ive exclusively, thereby diminishing the roles of the many other people responsible for the Zebra pen. It has been impossible to find either an original Orator or Zebra or any photos that show these objects other than the images of them to which the William Weaver Company owns the rights. Therefore, although both can be seen on the internet, neither is illustrated here.

Jony Ive took off for America at the invitation of Pitney Bowes to spend eight weeks at their headquarters in Stamford, Connecticut. He was not impressed. Everything he saw—the town, the products—was too staid and conservative for the tall young designer with his mullet cut shooting up like a porcupine's needles.

He decided to go straight from America's East Coast to the edge of the Pacific. The region on the southern side of San Francisco Bay had, for over a decade, borne the name of a crystalline solid grouped with carbon, tin, and lead on the periodic table, the second-most-abundant element in the earth's crust, with oxygen the only one of which there is more. "Silicon," a natural semiconductor, was not yet a word most people knew, but

"Silicon Valley" had become a hotbed of technological development, booming with new companies that used silicon chips in the integrated circuits essential to computers. Jony Ive yearned to be part of the future, and he knew this was where it was in the making.

5.

Jony Ive liked what he saw of California and the high-tech world during that glance of it in 1989. But he did not yet consider America a possible home base. The woman he had married two years earlier, Heather Pegg, awaited him at home in England, and he went back happily.

The young couple moved to London, where Ive joined Roberts Weaver Design. His impeccable draftsmanship helped make his innovative designs easy to realize, and a number of them were put into production. He continued to educate himself, delving into Dickens and other nineteenth-century novelists with an intellectual reach and a curiosity about human needs and wants that nourished his superlative design sensibility.

Ive also studied a range of approaches to the creation of household objects. The designers whose worked most swayed him included the Irish architect and furniture designer Eileen Gray; the leader of Italy's Memphis Group, Michele De Lucchi; the contemporary British designer Jasper Morrison; and Dieter Rams, a German designer who had brought Bauhaus sensibility to the household objects made by Braun, the highly successful manufacturing company that produced audio equipment, electric shavers, kitchen devices, and other household implements in Kronberg im Taunus, near Frankfurt.

Ive's taste was eclectic; his preferred designers differed vastly in style. But they shared the belief that what we use every day impacts our inner selves. Besides Gray's furniture and De Lucchi's playful and extravagant Memphis armchair, Ive particularly admired the frippery and futurist look brought to the world of luxury objects by firms such as Alessi, Artemide, and Philips. Makers of lamps and other household objects had previously been inclined to styles harking back to earlier eras; the designers for these European companies brought in a refreshing spark. Morrison's designs for everything from soup ladles to tram cars to plastic watering cans—minimalist yet stylish, with an individual twist that sets them apart—also delighted Ive.

But above all, it was Rams, born in 1932, who had a direct impact on Jony Ive. Rams had worldwide influence, primarily with the calculators, clocks, and audio components he made for the firm that Max Braun had started in Frankfurt in 1921. Its headquarters having been destroyed by Allied bombing at the end of World War II, Braun had quickly regrown as a maker of audio players and slide projectors. Of all the international brands, Braun produced, with Rams's understated and subtle aesthetic, the cleanest designs, possessed of rare logic and intelligence. Jony Ive's own work was in the Braun mold: absent any imposed "style," essentially functional and elegant.

6.

In 1992, Robert Brunner, head of industrial design for Apple Computer, asked the twenty-five-year-old Jonathan Ive to work on Apple's latest concept, a carefully guarded secret called Project Juggernaut.

Steve Jobs, having started Apple in 1976, had left the company in 1985. While Jobs was now building up the companies

Jony Ive admired the work of a range of industrial designers, from playful and outrageous to austere and refined, but the German Dieter Rams, born in 1932, had the greatest impact. Rams was the chief designer for Braun, a company based in Frankfurt; he gave their clocks and audio components an understated and subtle aesthetic. This audio system from 1956, with the logic and intelligence of its clear plastic lid covering components that needed to be dust-free, had a grace and functionality that Ive did his utmost to incorporate into his designs for Apple.

NeXT and Pixar, John Sculley ran Apple. Ive's design for Project Juggernaut never saw the light of day, but it was good enough to get him flown to Apple's headquarters in Cupertino. In the fall of 1992, Jony and Heather Ive moved to San Francisco so that he could become a designer for Apple.

Initially, Apple had had most of its products designed by an independent design firm run by Hartmut Esslinger. Esslinger had developed the concept of "Snow White" as a category for all of the company's products. Then the company developed an in-house design department, with Brunner as its head. It was not long before Ive assumed a major role in it. By 1994, his Newton MessagePad 110, the first Apple product manufactured in Taiwan out of parts made there, went into production. Then Ive developed the company's first flat-screen computer, a translucent laptop. His designs ranged from highly speculative ones to others that were commercially viable. Every day, he drove his orange Saab convertible the thirty-five miles to and from Cupertino and happily entered the precincts of high-tech culture while he and Heather were still based in the slightly more European milieu of San Francisco.

In 1996, John Sculley and Bob Brunner both left Apple. In 1997, Gilbert Amelio, who had been CEO for about eighteen months, was fired by the company's board of directors. Apple was losing money rapidly when the company acquired Steve Jobs's software company, NeXT, as part of a deal to bring Jobs back to run the company that had forced him out twelve years earlier.

In July of 1997, at a gathering of the board and key employees in the auditorium of Apple's headquarters, the interim CEO announced Jobs as the speaker. Unshaven, unkempt, dressed in shorts and sneakers, Jobs got straight to the point. What had gone wrong at Apple was "The products. The products suck! There's no sex in them anymore." Jony Ive, who was listening from the back of the auditorium, "wanted to quit." Jobs's coarseness, and his disrespect for computer models on which Ive had already worked, were hard to take. He imagined him and Heather moving back to England. But then Steve Jobs said that the goal of Apple was "not just to make money but to make great products."

Ive's immediate decision to stay would work out both for him individually and for the perpetuation of the Bauhaus values on which he had been nurtured. Steve Jobs felt that the issue of price should not predominate. The way to restore Apple to its former glory and take it beyond that into the stratosphere was through innovative, alluring design, even if it was expensive. Beauty and quality were the salient issues. He would rather create and produce luxury computers and related devices that were like alluring automobiles and fine wines than make anything second-rate.

Apple's products, by Jobs's calculations, would cost more than those being made by Dell and Compaq and the other companies that were taking the lead in the industry. They might not sell in the same quantity, but there could be a higher profit margin for something masterful; objects that looked truly great would turn Apple's finances around.

Steve Jobs still believed exactly what he had said in Aspen to that Bauhaus-savvy audience in 1983. The difference between the sublime and the ordinary was everything. Jonathan Ive was the person who would help him get there.

Part VIII

1.

Bauhaus design at its best demonstrates that the ordinary and the everyday, when addressed with discernment and tastefulness, have a candor and clarity that calm the insides. Nothing is false, nothing wasted. The sense of rightness directly benefits, if only to some small extent, the viewer's emotional state.

It would be specious to claim, however, that the iPhone necessarily has a positive impact on your inner state. Design is coupled with use. What you experience in your iPhone communications or the research the device facilitates is more likely to affect your feelings than will the appearance or touch of the object. The effect of using this taut and lovely tool may not be calming. It may well produce a fluttering heartbeat or whatever stressors your own metabolism induces.

But the hope of this book is not only that you will recognize the values that were evident in the house of Anni and Josef Albers—the modernism and open-mindedness that distinguishes creative people as disparate as Buckminster Fuller and Virginia Woolf and Vladimir Nabokov and Martha Graham, as well as the most perspicacious of the high-tech people—but that *you* will enjoy and celebrate those values, too. There are people who relish the sound of their dishwashers and washing machines, who prefer them to most "background music." It is a wonderful capacity.

It is the overarching ethos—the connection between appearance and inner workings, between surface and core moral values—that makes the Bauhaus, at its best, so vital. And in spite of all the annoyances, the irritation when things go wrong, the deliberate temptation to spend more money, that come with the use of the iPhone, many of the qualities realized in that small, sleek, tactilely and visually pleasant, handheld "telephone" that is also a computer and a storehouse of information and a family picture album and a source of music, have an impact very similar to Bauhaus objects at their best, as well as to those hardware stores that make practicality and integrity synonymous. There is a reigning thoughtfulness, and a delightful lack of fluff; what you do yields results.

Not that sheer functionalism is always a desirable choice. Huts in Africa are bare bones, but out of paucity of means, not because of a taste for lean aesthetics. In reverse, simplicity can be over-the-top. The expensive restaurant where a dollop of coriander mousse in the center of a vast white porcelain plate is supposed to be prized like a Giorgio Morandi still-life makes us crave franks and beans.

But bareness and visual austerity can serve and accentuate the capacity for effective action. They can induce the meditativeness of a monastery. Simple, versatile forms can be vehicles for hope and joy.

2.

Mies van der Rohe's "Less is more" encapsulates the Bauhaus aesthetic. So does his "God is in the details." But with Mies, those details were often ebony and travertine marble and fine pigskin. When the reductionism is applied to humble components, it is somehow more authentic. Josef Albers devoted his life to extolling the merits of "minimal means for maximum effect." In his basement workshop, he had two worktables: each a four-by-eight-foot piece of plywood resting flat on two sawhorses. Carefully positioned fluorescent tubes—arranged warm, cold, warm, cold—illuminated one table. Over the other, the tubes were installed cold, warm, warm, cold. Seeing paintings under both conditions enabled the artist to better understand the impact of individual colors on the colors adjacent to them. Josef's ingredients were nothing but unmixed pigments that he squeezed directly out of their tubes and applied with his painter's knife. Imagine two *Homages to the Square*, neither finished, both with the identical Reilly's Warm Gray 8 made by Winsor & Newton at the center and with the same Mars Yellow by Grumbacher in the square form that surrounds it. These two Masonite panels are the same size, and the measurements and proportions are identical, having been drawn in place lightly with a sharp pencil. Josef then explains to you that for the outermost square, which he is yet to make, he will apply his painter's knife "the way I spread butter on bread." In one case, he will use an Optical Warm Gray 3 by Shiva. In the other, he will use an Optical Cool Gray 3. It is imperative that it, too, is by Shiva. Optical Gray 3 by Blockx or a different manufacturer would be totally different in hue; Josef wants the only variable to be in color temperature.

The variable of "warm" and "cool" will cause totally different spatial movement and overall tonality between the two paintings. The inner two colors in both paintings look different, although they are actually the same. The only actual variation, in the outermost colors, with their distinction being only that one is cool and the other warm, transforms everything. Not only do identical colors offer different readings, but rhythm, position, and scale also seem to change. In his soft but emphatic voice, his sibilant *s*'s uttered with the conviction with which a calm priest declares what he knows to be a reality of existence, Josef declares this an example of "minimal means, maximum effect."

Of course the minimal means are only the appearance of minimal means. They are the by-product of technique developed through study, concentration, and intense experimentation.

The apparent simplification is the same as the operation of the iPhone. Just press down lightly in the right places and a great deal happens. The iPhone in its totality is like an Albers *Homage,* a seemingly lean and Spartan object that generates infinite events. With this single device you can travel for the day and achieve a vast multiplicity of tasks. Physically compact, deceptively plain in appearance: this is the essence not just of the Bauhaus, but of forms of wisdom that go back to ancient Greece. The Bauhauslers not only admitted but celebrated the qualities of the best classical objects. Modernity was not necessarily essential—although the latest and most current technology could be glorious. What counted was focus, paring down: the form of the Parthenon as opposed to all the ornament and complexities of wedding-cake architecture.

3.

Plutarch recounts the legend of Alexander the Great and the Greek philosopher Diogenes. While leaders from hither and yon call on Alexander to congratulate him on his ascendancy to power, Diogenes of Sinope chooses not to bother himself with the idolatry. Alexander reacts by going to call on the great thinker, who has opted instead to lie in leisure outside his house in the sunlight rather than render homage to the newly anointed emperor.

When Diogenes becomes aware that Alexander and his many minions are approaching, he props himself up slightly to acknowledge their presence. The monarch stands looking down at Diogenes. Deferentially, he asks the philosopher if there is anything he wants. Diogenes responds matter-of-factly, "Yes— stand a little out of my sun."

Alexander's cohorts are shocked by Diogenes's lack of reverence for their leader. But Alexander admires the audacious philosopher for his directness, his temerity, and his focus on the simplest of actions. To ask the monarch simply to move a single step reveals a perfect grasp of the connection between a tiny human movement and the greatest of cosmic forces: the sun itself. To link the minimal and the maximal is genius. To concentrate on the essential is to be effective.

Alexander tells his followers that if he were not Alexander, he would choose to be Diogenes.

4.

Of course the iPhone can also be a cold, irritating, nasty object. Its occasional hostility to its users further qualifies it as the

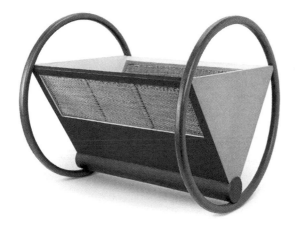

Peter Keler's "Bauhaus cradle," made in Weimar in 1922 and one of the first Bauhaus designs to become widely known, had a lot of the fundaments of the school's reigning aesthetics: bright and bold primary colors, painted flat, and the exclusive use of pure geometric forms like the circle, the triangle, and the rectangle. It was inventive and charming to look at. But who would put a baby in it?

emblematic Bauhaus object. Good design can be tyrannical. That famous Bauhaus cradle that Peter Keler made in 1922 is admirable for its simplicity. Its construction from lithe circular supports on each side, enabling rocking while supporting two planks of wood angled like a triangular gutter—sloping downwards in a V form to create the place where the baby would lie—is original and clever. But imagine putting a baby in there! How many pillows would be needed to make the infant comfortable? What would the risk of those pillows be to the child, who might roll over and suffocate? And wouldn't the pillows destroy the pure design that was the whole point anyway? Yes, the Kandinskyish coloring is delightful, and the object is fun to look at, but is it not indifferent to true human needs? Marcel Breuer's playpen is another travesty—if we consider the experience of a toddler rather than of an aesthetic purist. It resembles a miniature prison. The weaving workshops produced refined and dignified textiles that are modern masterpieces, but it also made some tapestries that resembles group projects made by nursery-school teachers or residents of a retirement community, with no element relating to any other, and the look of "artsiness" superseding any logic. Even the best of the Bauhaus designs can be undesirable at times. If you want to sit in a chair that affords sublime comfort while you concentrate on reading a book, the aesthetics of the seat—which you cannot even see, since you occupy it—matter less than its cushioning and its postural correctness. Wassily armchairs leave a lot to be desired for comfort and usability—the leather edging is unpleasant if you are wearing shorts; the back support should be lower; the seat begins to hurt if you stay there too long; your position has you too comfortable, too supine, if you are having a business meeting; it requires too much effort to get yourself up. The spirit and intentions of the Bauhaus were magnificent, but metric perfection and precise geometry of forms are not always all that is needed in life.

Still, nothing induces the same problems as these iPhones, except for computers themselves.

The iPhone, however remarkable, can seem willfully determined to get on your nerves. The problems begin with the packaging. All those pieces of shiny white cardboard give the impression that they are flawless and can be disassembled and reassembled with the flick of a wrist, but when you open the boxes you feel inadequate because you cannot get to your new toy without destroying the impeccably molded packing materials that contained it in transport.

The Apple logo insinuates itself into your field of consciousness, time and again, as if it considers itself God's gift to the earth. In the packaging, it has none of the grace or discretion it has on the actual object. This, too, is upsetting. Real apples are miraculous, but the abstract form of the fruit, with a bite missing, is altogether different. Entirely flat, lacking color or mass, it has none of the life of the fruit it represents. It has been grown in a design lab, rather than on a tree, and that smirking little leaf on its top is insufferable.

The iPhone can have a way of presumptuously insisting that it is better than you are. If you want to have it fixed or get an accessory for it, you are required to go into one of the horrid stores with metallic bad-breath air, and then inevitably wait for two hours after being told it would only be ten minutes. The salespeople act as if they have everything you need, but enjoy concealing it from you while you have to figure out what it is.

What has been built to serve you lords its power over you. The phone itself can intimidate its users; its marketing and servicing humiliate most everyone who is prey to it. Deliberate inconveniences are imposed alongside the advantages that make the device necessary, to ensure that you have to get help or spend more money.

One critique of the iPhone—and you could fill volumes with them—points out that one of the greatest boasts about its design is the way that it is devoid of buttons except for the single circular one at the bottom. Promotion from Apple explains that since you do everything on a touchscreen, not even the slightest muscular energy is required. But the result is that if something goes wrong with the screen, you are left with a useless object. In the days of the flip phone, there was still some hope even if the screen broke.

Smartphone wrath results in the regular appearance online of anonymous hissy fits. A highly imaginative explosion, aimed specifically at the iPhone, rants: "Apple loves to infantilize its users, imagining them to be a bunch of ironic mustache-wearing yupsters on the eternal umbilical to their parents, taking Instagram photos of the artistic foam in their latte. The idea of a person who might use their technology in an adverse situation is unthinkable because their target user probably hasn't even changed a tire." (While quoted repeatedly, this diatribe is never attributed to a source.) The device that has become central to everyday life has given birth to new sources of neurosis.

And then there are the problems in human relationships that never would have occurred without e-mail in general, with the "Reply All" function sometimes causing disaster. iPhones are often the vehicles of these crises. I once read a communication on my iPhone in which an art dealer with whom I was negotiating prices answered my e-mail increasing the cost of work by Josef Albers with an e-mail he thought he was sending only to a colleague to whom I had also addressed my communication. He wrote of me, "What a wanker!" The situation cascaded from there, in spite of my initial willingness to make light of it. Every one of you knows, I am sure, of mishaps in computer communication. The iPhone wins "Most Dangerous in Show" in this regard.

Part IX

1.

Don't think that our most high and noble art is taught or learned in schools or academics. What you discover there will be reworked as soon as you are able to observe forms and colors with love.

—Paul Cézanne to Georges Rouault, 1906

Cézanne was sixty-six when he wrote the letter to the thirty-five-year-old Rouault that contains the first epigraph to this book. The older artist would die within a few months of extolling the merits of direct observation of the essential elements of the visual world with an open heart. Living next to his studio in woodland that looks out toward Mont Sainte-Victoire on the outskirts of the small French city of Aix-en-Provence, he had no idea that an institution like the Bauhaus would ever come into existence. Art in France and Germany was still taught only with traditional methods. Exact representation mattered above all. There was a single way way of doing things, which depended, above all else, on repetitive drawing from live models.

Only two years after Cézanne's death, in 1908, when Josef Albers was twenty, he saw Cézanne's work for the first time. Karl Ernst Osthaus, a brave connoisseur of recent innovative art that few people liked and even fewer bought, had created a museum called the Folkwang in the German town of Hagen. For the rest of his life, Albers considered his viewing of the two Cézannes at the Folkwang "the start of everything." Cézanne reduced form to its essence and unleashed color. He used empty space magically. Albers was just one of the guiding lights at the Bauhaus who worshipped Cézanne. The Bauhaus was an attitude, not a style or a time period.

Personal engagement, open eyes, and receptivity come only from contemplation. Cézanne was the essence of Bauhaus methodology. Throw in technical savvy, and Cézanne's counsel to Rouault would apply to the approach of Steve Jobs. Observation precedes vision. The sort of accumulated knowledge of useless facts that leads to high grades on exams is beside the point. Experimentation in the making of a product, and the invitation for further experimentation once the product exists, and the emphasis on pleasure, are what matter.

2.

Josef Albers wrote a poem about Cézanne that emulates its subject in its lean, elusive beauty.

CÉZANNE

You see
but you do not
see

But
later
that much
is the same

as one
and the same
is many

Far later
the wonder
that the same
is all

This homily by the teacher of the Bauhaus foundation course and master practitioner in its glass, metal, carpentry, graphics, and photography workshops takes us straight to the iPhone. Albers's "You see but you do not see"—the flux between what is constant and what is always in motion, the simultaneous presence of the singular and the infinite—is the magic permitted by that compact little instrument in your hand. No one would claim that the iPhone is possessed of the earthy majesty of a Cézanne landscape, but the flux between the specific and the amorphous, established thanks to meticulous craftsmanship, is there.

3.

In the Bauhaus workshops, you learned technique and you mastered your materials. And then you went your own way, without a prescribed, single way of doing things. You might even attend Paul Klee's lectures on the flow of water starting in rivulets at high altitude and continuing downward into small mountain brooks into rushing streams into wide rivers, and from there being absorbed by the ever-fascinating sea, as exemplary of the splendor of earthly life and of the development of form. But it was not because you were expected to reproduce those events. Rather, you were encouraged to make them part of your understanding of earthly existence as well as of progressive processes. Once you had a solid base and sufficient knowledge, and respect for the essential facts and real needs of life, nothing rivaled careful observation and contemplation fired by a passion for the invention of new means for experiencing life.

For most of the twentieth century, America was obsessed with a very singular notion of education completely at odds with the searching and exploration that the Bauhaus endorsed. The reigning beliefs about learning in relation to achievement in prestigious universities started in the original thirteen colonies. The world in which Steve Jobs triumphed considered degrees from Ivy League colleges and their equivalent fundamental to leadership and fortune. Jobs prevailed spectacularly outside that tradition. His lust for venturing into the unknown and his will to go beyond the constraints with which most people imprison themselves were Bauhaus in spirit.

Jobs's formal education was desultory at best. The Jobs family lived in Mountain View, California, a town south of Palo Alto. Jobs's father worked in Palo Alto but could not afford to live there or in its expensive suburbs, where the public schools were superior to those in Mountain View. Jobs's mother, a bookkeeper, taught her son to read before he went to kindergarten, thereby minimizing the liability of his dyslexia, but nothing compensated for the dull education at Monta Loma Elementary School, and young Steve basically hated the place. "I was kind of bored . . . so I occupied myself by getting into trouble. . . . I encountered authority of a different kind than I ever encountered before, and I did not like it. They came close to really beating any curiosity out of me."

For the most part, Jobs's education did not improve after elementary school. His high school was no match for America's chosen training ground for leaders, the private schools and colleges with glossy reputations to which families like the Fords and the Rockefellers send their children; and he lacked the motivation or academic abilities that enable rare individuals to flourish at large public schools. Few people would have imagined that someone with Steve Jobs's background could be so much more successful, by every measure, than the people taught to study and cram and memorize in formation to acquire the patina of education that is supposed to lead to successful jobs and financial prosperity. Yet given the impact of both Bauhaus designs and Jobs's contributions to society, it is clear that Cézanne's traditional "schools or academies" have never been, and still are not, an ensurance of learning that matters. Experimentation, combined perhaps with a bit of discomfort, is infinitely more effective.

4.

The Bauhaus was itself a school that was unlike any other. By picking Dessau, an industrial city that, in spite of its nice park, did not have much to offer except for being not too far from more charming places like Leipzig and Berlin, it was declaring a form of independence from usual notions of prettiness and from reverence for historicism. Its masters were on the fence about how hierarchical to be; they debated, time and again, whether or not they wanted to be addressed as "Professor." But they did not necessarily have any academic credentials.

The assumptions in mainstream German society were the same as in America, consistent with beliefs going back for centuries. Successful professionals not only needed first-rate university education, but were expected to go on for additional years of academic work to become doctors, lawyers, aeronautics engineers, diplomats, or any other kind of professionals. No one thought art schools were in the same echelon as other specialized schools. Nor did people yet entertain the idea of training to be a professional athlete. Both the focus on manual skills and the deliberate absence of a traditional school curriculum in Weimar and Dessau were revolutionary. The iPhone was invented by someone whose education was as nonstandard as everything about the Bauhaus. And like the products of the Bauhaus and the life of the school, this handy object has a freshness and immediacy, a spark to it, for precisely that reason.

Students at the Bauhaus came from every walk of life. While Kandinsky descended from Russian nobility, Breuer was a dentist's son. Few had working mothers. Jobs was born to an automobile mechanic and a bookkeeper. There is no formula for what gives people gumption and imagination and brilliance. What is certain, though, is that, to achieve the willpower and capacity to dare new things, you cannot depend on your heritage.

People like Steve Jobs and the Bauhauslers used their hours differently from most people consciously seeking success. Instead of doing what was requisite for taking the next step to the higher rungs on the ladder, they experimented. They thought; they imagined; they failed; they succeeded.

Jobs and the Bauhauslers made miracles happen because they recognized the inherent flaws of a lot that was regarded as sancrosanct. Schlemmer's costumes for his *Triadic Ballet,* and Kandinsky's sets for *Pictures at an Exhibition,* deliberately countermand their dreary precedents. The iPhone may have certain purposes of old telephones, but it provides a conscious contrast between what is pocket-size, compact, and fast and what was large and cumbersome and slow. Playfulness, insouciance, originality, and freshness are palpable in these objects. They are like joyous revolutionary chants. The rebellion of their creation animated them.

The iPhone—and the design aesthetic of the Bauhaus—make manifest the freer approach to education that in the eyes of the larger public initially violated the good old traditions of "the groves of academe." Mary McCarthy gave a novel that title to suggest the internecine, time-and-energy-wasting infighting within the bastions of the higher tiers of liberal education in America. It comes from Horace's writing about the search for ultimate "truth" in the inner sanctums of educational pursuit. The highly touted professors McCarthy acutely deprecates manipulate the system to their own benefit. They are more interested in advancing their careers than in seeking knowledge that might advance society at large.

Jobs and the doyens of the Bauhaus had visions that corresponded to Horace's ideals, and stuck to them without wavering. They never let side issues dominate their creative course. The iPhone, like Anni Albers's open-weave textiles and Paul Klee's paintings, has a weightlessness and the lack of encumbrance with which Jobs and the Bauhauslers tried to live. And it shares the same spark of invention.

That last quality cannot be learned; nor can it be fully explained. But it brightens life immeasurably.

Mies van der Rohe had learned rudimentary construction technique from his father. Josef Albers said that "all that counted" in his own education was the practical know-how inculcated in him by his own father, especially in painting and woodworking. The first thing young Josef painted was wooden crosses on graves, with his father demonstrating the technique of brushwork. Albers told the many scholars eager to identify assorted aesthetic influences or to associate him with Yale, when he headed its Department of Design, or with Harvard, when he lectured there, "I came from Adam and my father; that's all."

He considered reference to the academic style practiced by one's teachers or to the "zeitgeist" of the moment a travesty. Steve Jobs's father worked for a finance company, at a low-level job, but he was a trained car mechanic and loved building things. He had briefly been a used-car salesman; anything to do with automobiles beckoned him, even though his preference was for the details of their inner workings. When Paul Jobs constructed a stockade-style wooden fence around the family's yard, he gave his son a hammer and taught him how to hit the nail correctly.

Steve did not share Paul's interest in fixing old cars, however. His father would later say that Steve did not like getting his hands dirty. In a father-son relationship with the father a mechanic, the comment suggests a world of differences. But Steve liked what he could learn about electronics when he joined his father in the garage.

5.

These perfectionists obsessed similarly about the surfaces of the objects they created. The man who brought Apple products and the iPhone into the world, like the designers of Barcelona chairs and clear glass lighting fixtures and simplified textiles celebrating new materials and construction techniques, carefully orchestrated their presentation. Like them, he measured the impact of the details while simultaneously being certain not to wear his heart on his sleeve.

The Bauhauslers' fastidious appearance and Jobs's inevitable sweatshirt and jeans and sneakers share the awareness that governs them in spite of the contrast of a polished surface and a rough one. These insightful people knew the impact of details. They considered and reconsidered their received personas as they did the objects they put out into the world. They carefully masked their private selves with strong veneers.

Jobs was initially well groomed and neatly dressed in public. The younger Steve Jobs often wore jeans, but they were immaculate, and his straight black hair was always trimmed meticulously and carefully brushed. He was clean-shaven. In the early years, he often appeared in a well-cut suit and tie. In his wedding photos from 1991, his formal bow tie is perhaps a centimeter off center on top of the wing collar of his formal shirt, but his dinner jacket is buttoned carefully and fits perfectly. But as Jobs became more successful, he got deliberately schlumpy in public—seeming not to care, as if other issues were so vital that he had no time to worry about his appearance. He was no longer clean-shaven, nor was he neatly bearded. His wrinkled shirts were half tucked in, half hanging free of his equally wrinkled shorts or trousers. It was only with great success that the slobbiness came in.

Paul and Lily Klee wore dashing and elegant white linen resort wear on holidays. In Weimar, the Bauhaus faculty went out of their way to look like good solid bourgeoisie. The Bauhauslers were, for the most part, the sort of Europeans who dressed up to do everyday errands. The men wore suits, the women dresses. Kandinsky sometimes painted in formal attire. None of this was accidental; the people who presented themselves as correct bourgeoisie were deliberately differentiating themselves from the bohemian types like Johannes Itten. They believed that they would be more effective advancing revolution if they lived with manners and form. The Bauhauslers maintained a code of politeness and correctness. They genuinely loved measure and rules, in language as in comportment.

Steve Jobs was from an entirely different culture. He was in many ways quintessentially American. Yet that qualifies him all the more as an exemplar of Bauhaus beliefs. That he produced the quintessential Bauhaus object is evidence that the ideals of the school were as they purported to be—capable of transcending all that was European, going beyond what was unique to a single culture, and spreading across the map of the entire world. Part of the success of the revolutionary objects produced in the Dessau workshops and at Apple Computers is that the figureheads associated with them appeared unthreatening and in keeping with the places and times with which they lived. But underneath their seemingly ordinary costuming, they addressed what was universal and timeless.

Steve Jobs made it through his high-school years with marijuana playing no small part; LSD had its role as well. He then went to Reed College, in Portland, Oregon. One of the most progressive small colleges in the United States, Reed eschewed mainstream requirements and gave its students rare freedom. Still, Jobs did not stay the full four years. Lisa Brennan-Jobs's book has him there for only a single term, Isaacson's for a couple of years. Regardless, Jobs did not make it to graduation even in a countercultural institution.

Johannes Itten had scandalized Weimar society because he and the fellow adherents of Zoroastrianism cavorted in the nude in the public park and could be seen doing their ritual pinpricking to let the toxins out of the systems they had just nourished with fruit juices with laxative effects.

The other Bauhauslers, and Steve Jobs, knew better than to shock with their surface appearance even if everything else about them was the essence of revolution. They sometimes followed the rule book, but only because they consciously chose to. They were not coerced; they made their choices attentive to their impact—with their eyes wide open and their originality inviolable.

Part X

1.

The year after "Apple Computer" was incorporated and the name made official, its trademark logo was put to use. The style of the logo, as well as the missing bite, tailored the familiar piece of fruit to its new purpose.

This was when Steve Jobs and Steve Wozniak hired Regis McKenna to help them have their new products known broadly and in the style that suited them.

McKenna in turn hired Rob Janoff, a young California-born graphic designer, as his art director, and Janoff designed the logo.

To develop the logo, Janoff initially bought a bag of apples and cut them up. Jobs insisted that the logo should have color in it, overruling McKenna's view that the printing costs would be prohibitive. Janoff presented to Jobs two versions of the piece of fruit with a rainbow of horizontal stripes across it. One was a whole apple; one had a bite missing. Jobs picked the latter, saying that otherwise the fruit could be mistaken for a cherry.

One of the reasons the iPhone is more true to Bauhaus values than many other Apple products has to do with the style of the apple. When it has bright stripes across it, as it does on the Apple II computer, it is garish. The stack of vulgarly colored rectangles represents a level of decoration few people at the Bauhaus would ever have condoned. It is only when the apple gets simplified to an austere, minimalist representation of the fruit that we are brought to the essentiality that was central to the highest standards of the Bauhaus.

There is a popular myth that the apple with a bite missing is an homage to the British inventor Alan Turing. Turing was one of the seminal figures in the development of computers. His life ended tragically, not only because so little of what he did was acknowledged at the time, but also because he was vilified for his homosexuality. When he was about to serve time in prison for what was then illegal in England, he killed himself by eating an apple that he had laced with cyanide. But Rob Janoff says that, appealing as it would be to consider the apple a tribute to Turing, there is no truth to the story.

When asked, Steve Jobs simply remained silent about the notion of the Turing reference. This was in character. He delighted in ambiguity about all of his important marketing decisions. His general policy was to stay mute about what had led to what. His phone call to the developer of Gorilla Glass to shower thanks the day the iPhone was released was exceptional.

2.

Apple Inc. is vigilant about how and when the logo is used, with its precise dimensions and the size of the missing bite and the abstracted leaf on top copyrighted and protected accordingly.

The bite missing on its right side gives the logo the imprint of a smile. The single leaf dancing above it is jaunty—a great touch except when it becomes irritating in packaging materials.

The Apple logo, plain and simple, is a masterpiece. In subtle relief, absent color, or as a bare outline, it is voluptuous in form, jaunty in spirit, and mischievous with its missing bite. But to many viewers, once it bears rainbow stripes in garish Day-Glo colors, it is tasteless and irritating. While the outrageous Edsel and cumbersome Lisa had a certain charm, the apple in loud pajamas violates the classicism of the shape left without decoration.

The wavy bottom makes the single piece of fruit both stable and fluid.

This logo has been thought through in every way: the proportions, the outline, the lack of surface variation. Take away the rainbow of nursery school colors, and it embodies Bauhaus precision and care. In outline form, it fulfills the Bauhaus goal of visual entities with widespread impact. Its effect is not just because of what it represents but because of its curves, its balance as well as deliberate imbalance of linear movement, of millimeter-perfect spacing between elements, of points in opposition to lines. Kandinsky in particular taught this relationship of point and line; Klee in his teaching emphasized the progression inherent to natural growth. Part of the Bauhaus belief system was to make creations comprehensible to the large audience, and the name Apple and the logo representing it fulfilled those purposes.

Apples themselves, however, have different associations for different people. To some, an apple is, in the era of large supermarkets that ship comestibles from one end of the country to another in large trucks, a by-product of dangerous chemical engineering. The modern commercial apple has undergone a range of unknowable processes and had multiple potentially harmful substances applied to its surface. An equally vast number of artificial compounds has been employed in its nurturing. Wrapped tightly in plastic, made seemingly ageless, it might be on sale at a magazine stand, looking not much more edible than the printed publications alongside it. But that is a cynic's version of the apple.

An apple can also be seen as a miraculous fruit that has just fallen from a tree, or been gently removed because it has reached its prime and can break free easily. The healthiest specimens have been grown organically. Ideally, they are nurtured in climates where they go from blossom to fruit in the warmth of spring and summer, and are at their prime with seasonal change in the autumn. They should be picked or collected from the ground shortly after the first frost of the year. It takes that jolt of cold temperature to provide the snap, the tactile qualities of crunchiness, as opposed to mushiness and too little fight to the bite. Apples harvested at the right time have fantastic flavor and provide one of the most thrilling culinary pleasures imaginable. Eating one is a direct encounter with a miracle.

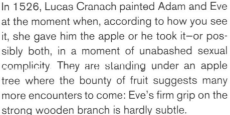

In 1526, Lucas Cranach painted Adam and Eve at the moment when, according to how you see it, she gave him the apple or he took it—or possibly both, in a moment of unabashed sexual complicity. They are standing under an apple tree where the bounty of fruit suggests many more encounters to come: Eve's firm grip on the strong wooden branch is hardly subtle.

Albrecht Dürer painted this *Eve* in 1507. The apple, spotlighted like his subject's breasts, is as unabashedly sensuous as the woman holding it.

Art historians and theologians, however, can see the apple as the forbidden fruit Eve offered to Adam. The apple is the central object in the biblical tale that makes desire and sexual pleasure sources of guilt. The scene of Eve offering it and Adam taking it has been the subject of a lot of the world's greatest art. Whether Eve's beckoning or Adam's accepting is the more significant of

When Masaccio painted the expulsion from paradise between 1423 and 1427 in his ravishing frescoes in Florence's Brancacci Chapel, he made the consequences of consummated lust—symbolized by the bite from the apple—totally tragic. Adam and Eve are devastated by shame and regret. The missing bite from the eponymous fruit of Apple Computer was intended to suggest no such anguish—however painful the consequences of using an iPhone or personal computer might be.

the two acts—or whether both players are equally responsible—depends on the artist. Dürer renders the scene with the most seductive of Eves, her flesh and the allure of the apple she holds out utterly glorious; she is the one in charge. Romanesque church portals, with their imagery compressed into an arch above and side panels alongside the door that will lead people into the world where they seek redemption, similarly show Adam as weak: as if he cowers, theoretically unwilling yet still compliant, in response to Eve's offer of sexual pleasure. But in other artworks Adam responds with blatant lust and an enthusiasm equal to Eve's.

Lucas Cranach the Elder shows Adam and Eve with that apple linking them exquisitely, the pair complicit in their taking of pleasure, the impending glory of their physical union so halcyon that most of us would gladly accept the expulsion from paradise as its price. Masaccio and Masolino, however, illustrate, in the Brancacci Chapel in Florence, that exile from paradise so excruciatingly, and with painting so convincing in its conjuring of physical flight and emotional anguish, that one wishes Adam had said no; both of the lovers are victims of their desire.

3.

What the bite out of the apple meant to Steve Jobs personally, perhaps unconsciously, may have had something to do with his own life experience.

The birth certificate of the future Steven Paul Jobs names his mother as Joanne Schieble. Schieble's father had immigrated from Germany to northern Wisconsin, part of a large influx of people who made that same journey from central Europe to the American heartland. The future Steve Jobs's father was Abdulfattah Jandali, a Syrian Muslim who was a teaching assistant at the University of Wisconsin. Schieble and Jandali were not married.

At the Bauhaus, such nonstandard family situations had been accepted in a way they were not in Wisconsin in the 1950s. Walter Gropius was bringing up Alma and Gustav Mahler's child, the little girl whose grandmother had kept her from seeing what was going on when Alma and Gropius were cavorting at the sanitarium, and treated the child as if she were his own biological daughter. Then, in his first year of running the Bauhaus, he and Alma had a baby. Or so Gropius thought. Two weeks after the birth of the little boy he considered to be his son, Gropius overheard Alma on the phone with the writer Franz Werfel, and realized that Werfel was the actual father. Gropius's main reaction was to make sure that nothing around this revelation upset Alma when she was nursing; drama would be bad for both the mother and the baby.

Joanne Schieble's father was not so enlightened or free-spirited. Arthur Schieble, who owned a mink farm near Green Bay and worked in other businesses as well, threatened to disown his daughter when she fell in love with the Muslim Jandali. Still, Joanne spent the summer of 1954 with her boyfriend in Syria, and by the time the two returned to Wisconsin, she knew she was pregnant. Joanne's father, who was dying, said he would cut her off completely if she married Jandali. Before the child was born to these unmarried young lovers, Joanne had taken refuge with a doctor in San Francisco who agreed to deliver the baby and arrange an adoption. So there was never any possibility that their son would bear their names.

Paul Reinhold Jobs was also from Wisconsin. A high-school dropout, Paul had made a scrappy living as a mechanic until he joined the Coast Guard. During World War II, he served on the boats that took troops to Italy for General Patton. When the last ship he was on was decommissioned in San Francisco after the armistice, he wagered a bet with one of his friends that he would find a wife within two weeks. Clara Hagopian, from an Armenian family that had fled the Turks, had been brought up in New Jersey, but her family had moved to San Francisco. Paul won his bet.

The worst hurdle in Paul and Clara's marriage was when Clara had an ectopic pregnancy. It left her unable to become pregnant again. Nine years after marrying, they decided to adopt.

Steve Jobs's relationship to his daughter Lisa, the child born out of wedlock, was ambivalent at best, having begun with complete denial. All that is certain is that his explanation that the bite missing from the apple is there to differentiate the fruit from a cherry was probably his way of deflecting the truth. That bite is the ultimate symbol of sexual lust—and its possible consequences. And to users of Apple products, the bite suggests: "Partake!"

Go ahead—use this communications device to line up a hot date, to look at pornography, to enter forbidden territory!

While the public acknowledgment of erotic joy is still taboo for many people, Jobs and his cohorts gauged correctly in thinking that the naughtiness of the missing bite was too subtle a way of evoking the pleasures of sex to raise hackles. And in this case, with the story of Adam and Eve being conjured, the interaction was the most universally acceptable one: monogamous and heterosexual coupling—hardly a crime in the world of a Reed College student, even if it was premarital. And, after all, knowing he had been born to unmarried parents, Jobs was unlikely to condemn their lovemaking, even if it was unsanctioned like Adam and Eve's and caused his maternal grandfather to insist that the baby go elsewhere. Rather, in spite of having initially denied that he was Lisa's father, Jobs was inclined to celebrate the societal transgression that led to his own existence.

Part XI

1.

Paul Cézanne imbues apples with an extraordinary presence. He manages through the painting technique he invented to capture the solidity and mass of this fruit. Throughout his life, he returned to apples intermittently. Universal and familiar, they suited him more than rarified subjects did.

Cézanne's canvases and sheets of fragile watercolor paper are evidence of the artist's capacity to evoke three-dimensionality regardless of the flatness of his medium. The invisible sides of the apples are palpable. Each single piece of fruit effectively displaces the air it occupies. These paintings make real the flesh as well as the skin, the way apples grow, the cores we cannot see, the taste, the nourishment. The art is so convincing that it conjures the trees on which in an earlier stage these apples were once only pretty blossoms in springtime. Cézanne's robust brushstrokes, their brutal honesty, their toughness, and his choice of colors, constitute truth.

2.

The art historian Meyer Schapiro's observations about Cézanne's apples elucidate the reason that the name and the logo Steve Jobs gave his company are part of the reason that the iPhone strikes an important chord within so many millions of people. Schapiro's text makes clear the links between what might seem to be the most specialized, seemingly irrelevant study of fine art—and the world of privilege of those people who have access to the art of Paul Cézanne—and issues that pertain to everyone. The renowned professor's close examination of the values and passions of the elusive Cézanne explores the depths of art making and the motivations behind the choice of subject matter. In so doing, it helps us better to understand the appeal of imagery that constitutes the daily visual bread of humankind and that Jobs cannily selected.

This step from refined intellectual pursuit to everyday global existence was the precise objective of the Bauhaus. The motives for the creation of the school are further realized in the apple that Steve Jobs picked as a name and a team of skillful designers refined as a logo. Apple's apple has now become part of a universal language better known than any verbal or written tongue. This was what the Bauhaus sought: for careful consideration of details to engender results that enter the mainstream of human existence. The Bauhaus wanted not just to render the visible and the touchable as sources of everyday pleasure, but to add the sort of spark that comes with an apple.

Cézanne's apples, of course, are imperfect in form. Some are ripe, others not far from rotting. Their blemishes are integral to their truthfulness. The apple that is the logo for all the products of the company so named is more of caricature. Perfectly round, it has a flat surface and an impeccable, thin outline. It represents the fruit without any of its realities of weight, texture, growth, or disintegration.

This corporate apple has an artificial resilience that functions as advertising. This apple will not rot; it does not have any flaws. The missing bite is a sure sign that consumption has begun, but nothing will change from here on. The suggestion is that your iPhone, or your computer mouse, or your laptop computer will be the same: ageless, pristine forever, permanently in ideal condition.

That deliberate illusion reflects sharp business tactics, the sheer commercialism of people who sell things, not the real-

To Paul Cézanne, apples were an unequaled source of richness—as an object to paint and a symbol of life's bounty. In this 1880 self-portrait, Cézanne's bald forehead has an uncanny resemblance to the fruit he prized.

ity of apples themselves or of their rendition by Cézanne. But what Meyer Schapiro points out as the essence of apples in all people's thinking and desires enhances our understanding of the motif Jobs picked for his company, even if Jobs then used it to foster the myth that nothing but good times lie ahead.

3.

Meyer Schapiro starts with Cézanne's early mythological paintings. In a *Judgment of Paris*, Paris presents "an armful" of apples to the winning nude female; apples are linked to female sexuality. Schapiro then quotes a letter in which Paul Gauguin

Cézanne's 1880 *Judgment of Paris* shows Paris handing over a bounteous bowl of apples. The centerpiece and focal point of the complex canvas, the ripe fruit is a perfect symbol of the attributes on Paris's mind. The richness of the simple apple, and all the implications of the fruit, make it likewise a perfect symbol for the iPhone and the rest of Apple Computer's communication products.

describes Cézanne: "A man of the south, he spends days on the mountain-tops reading Virgil and gazing at the sky." Then Schapiro introduces the Latin elegiac poet Propertius:

> In one of his elegies . . . Propertius chants the love of a girl won by ten apples. Addressing Virgil, whom he hails as greater than Homer, the poet writes: . . .
> *You sing beneath the pines of shady Galaesus*
> *Of Thyrsis and Daphnis with the well-worn pipes;*
> *And how ten apples can seduce a girl*
> . . . .
> *Happy man, who can buy love cheaply with apples!*

Schapiro will show that for Cézanne, apples represented acts of giving as expressions of love.

Using the fruit he saw in such quantity at All One Farm as the logo of his burgeoning computer company, Steve Jobs appreciated its association with bounty and with acts of generosity.

When we run our fingers over the apple on the back of an iPhone, it is the one smooth and shiny element on the anodized aluminum casing. The apple is like a small round mirror set in a large plaster wall. With its impact maximized by minimal presentation, its design universal and timeless, issues like individual authorship or financial value are out of the picture. The compact form with its silky feel penetrates the unconscious of the iPhone user as a suggestion of generosity.

Schapiro tells a story of Cézanne and his close friend Émile Zola:

> Cézanne could more readily respond to this classic pastoral theme, since in his own youth a gift of apples had indeed been a sign of love. In his later years he recalled in conversation

that an offering of apples had sealed his great friendship with Zola. At school in Aix Cézanne had shown his sympathy for the younger boy who had been ostracized by his fellow-students. Himself impulsive and refractory, Cézanne took a thrashing from the others for defying them and talking to Zola. "The next day he brought me a big basket of apples. 'Ah, Cézanne's apples!' he said, with a playful wink, 'they go far back.'"

In telling this story to his admirer, the young Aixois poet Joachim Gasquet, the son of a schoolmate and friend of the painter, Cézanne was not only joking about the origin of a theme of his pictures. He was remembering the painful rupture of his long friendship with Zola that followed the revelation, in the latter's novel *L'Oeuvre* (1886), of Zola's view of his old friend as a painter *raté*. Cézanne, in recounting the golden episode of their youth, is saying: There was a time when Zola could think of no finer expression of gratitude and friendship than a gift of apples; but now he rejects my apples.

Twenty years before, Zola, in dedicating to Cézanne his brilliant first venture into art criticism, *Mon Salon*, had written: *"Tu es toute ma jeunesse, je te trouve mêlé à chacune de mes joies, à chacune de mes souffrances."*

At an age when into the friendship of boys is channeled much of their obstructed feeling for the other sex, Cézanne confided to his closest companion, Zola, his poems of erotic fancy. But he shyly withheld his translation of Virgil's second eclogue on the shepherd Corydon's love of the boy Alexis. *"Pourquoi ne me l'envoies-tu pas?"* wrote Zola. *"Dieu merci, je ne suis pas une jeune fille, et ne me scandaliserai pas."*

As Propertius converted the theme of the offering of

apples in Virgil's third eclogue to one of heterosexual love, so one may regard Cézanne's picture as a transposition of that boyhood episode with Zola to his own dream of love.

Apples are sexual in many ways. Schapiro amplifies on Cézanne's regular return to them:

The central place given to the apples in a theme of love invites a question about the emotional ground of his frequent painting of apples. Does not the association here of fruit and nudity permit us to interpret Cézanne's habitual choice of still-life—which means, of course, the apples—as a displaced erotic interest?

One can entertain more readily the idea of links between the painting of apples and sexual fantasy since in Western folklore, poetry, myth, language and religion, the apple has a familiar erotic sense. It is a symbol of love, an attribute of Venus and a ritual object in marriage ceremonies. *Fructus*— the word for fruit in Latin—retained from its source, the verb *fruor,* the original meaning of satisfaction, enjoyment, delight. Through its attractive body, beautiful in color, texture and form, by its appeal to all the senses and promise of physical pleasure, the fruit is a natural analogue of ripe human beauty.

In Zola's novel, *Le Ventre de Paris* (1873), which has been called *"une gigantesque nature-morte"*—the story is set in the great food market of Paris, the Halles Centrales—fruit is described in a frankly erotic prose. *"Les pommes, les poires s'empilaient, avec des régularités d'architecture, faisant des pyramides, montrant des rougeurs de seins naissants, des épaules et des hanches dorées, toute une nudité discrète, au milieu des brins de fougère.*" The young woman who presides over the fruit is intoxicated by its fragrance and in turn transmits to the apples and pears something of her own sensual nature. "*C'était elle, c'étaient ses bras, c'était son cou, qui donnaient à ses fruits cette vie amoureuse, cette tiédeur satinée de femme . . . Elle faisait de son étalage une grande volupté nue . . . Ses ardeurs de belle fille mettaient en rut ces fruits de la terre, toutes ces semences, dont les amours s'achevaient sur un lit de feuilles . . . au fond des alcôves tendues de mousse des petits paniers.*"

In associating the woman and the fruit in this lush description, Zola follows an old *topos* of classic and Renaissance poetry. In pastoral verse since Theocritus the apples are both an offering of love and a metaphor of the woman's breasts. In Tasso's *Aminta* the satyr laments: "Alas, when I offer you

When Cézanne painted a plaster Cupid in 1894, he surrounded the symbol of love with a panoply of apples. That ordinary fruit was sufficient to suggest all that the mythological son of the war god, Mars, and the love goddess, Venus, represented. In Greek, the word for the Latin-derived "Cupid" is "Eros." He is the god of physical attraction, erotic feelings, and desire—presumably fulfilled. All those possibilities are suggested by the symbol of the iPhone.

beautiful apples you refuse them disdainfully, because you have more beautiful ones in your bosom."

The classic association of the apple and love has been fixed for later art, including Cézanne's, through the paraphrase by Philostratus, a Greek writer of about 200 A.D., of a painting of Cupids gathering apples in a garden of Venus. The Cupids have laid on the grass their mantles of countless colors. Some gather apples in baskets—apples golden, yellow and red; others dance, wrestle, leap, run, hunt a hare, play ball with the fruit and practice archery, aiming at each other. In the distance is a shrine or rock sacred to the goddess of love. The Cupids bring her the first-fruits of the apple trees. . . .

One may suppose that in Cézanne's habitual representation of the apples as a theme by itself there is a latent erotic sense, an unconscious symbolizing of a repressed desire.

4.

Meyer Schapiro also emphasizes the lack of anecdote in apples as subject matter, the wonderful ordinariness. He likens them to pebbles: "a pebble could serve as a sufficient theme in painting."

Pebbles! These were adulated by true Bauhauslers, for their visual wonder and for the history with which they were made. Paul Klee painted them. Josef Albers wrote poetry about them:

> *Easy—to know*
> *that diamonds—are precious*
>
> *good—to learn*
> *that rubies—have depth*

> *but more—to see*
> *that pebbles—are miraculous*

Of a pebble, and of an apple, Schapiro writes, "At first commonplace in appearance, it may become in the course of that contemplation, a mystery, a source of metaphysical wonder." Schapiro treats the apple not in its associational value in the story of Adam and Eve, but as something more generic—"completely secular and stripped of all conventional symbolism, the still-life object," and, in that capacity, as "the meeting-point of boundless forces of atmosphere and light." That apple on the back of the iPhone—with its shimmering surface that, in the otherwise

In this late still-life, dated 1895–1900, Cézanne reduced his visual vocabulary, just as his followers at the Bauhaus would. He gave two apples—one of which has the jaunty single leaf of the computer 'company's logo—unprecedented power and forcefulness. These are the qualities that buyers of Apple products are supposed to feel they are acquiring with their iPhones or other personal computers.

matte panel, causes it to reflect sunshine or lamplight or whatever else may bounce off of it—while as ordinary as a pebble, is as miraculous. Meyer Schapiro also sees Cézanne's apples as being like pebbles in their humility—their disassociation from issues of status or class. As such, Cézanne's apples represent something essential to the artist's sense of himself.

The apple was a congenial object, a fruit which in the gamut of nature's products attracted him through its analogies to what he felt was his own native being. In reading the accounts of Cézanne by his friends, I cannot help thinking that in his preference for the still-life of apples—firm, compact, centered organic objects of a commonplace yet subtle beauty, set on a plain table with the unsmoothed cloth ridged and hollowed like a mountain—there is an acknowledged kinship of the painter and his objects, an avowal of a gifted withdrawn man who is more at home with the peasants and landscape of his province than with its upper class and their sapless culture. This felt affinity, apart from any resemblance to his bald head, explains perhaps the impulse to represent an isolated apple beside a drawing of himself.

Not that Steve Jobs's artistry is comparable to Cézanne's, and it depends on your own viewpoint as to which of the two made a greater contribution to humankind. And while for Jobs there was no issue of baldness, it is similarly significant that the symbol he chose was, for him personally, associated with a hippie commune. This man who became such a part of corporate America, of the world of Wall Street and stock prices and corporate takeovers and capitalism at its most excessive, had made the central image of his life one he associated with informality, sharing, smoking dope, a certain carefree, devil-may-care, joyous revelry, the world of people who would rather dance to the summer solstice naked than attend stockholders' meetings to which CEOs have flown in private planes. The life of All One Farm was in his soul.

"All One Farm." The ambiguous name suggests a lot, including free love. Using its singular product as the name of his company, and its simplest rendition as its symbol, was not merely a way to inspire unconscious thoughts of sexuality in a vast client base, but a means of suggesting sex without embarrassment. Jobs could, after all, never have used a more specific sexual theme.

Meyer Schapiro's Cézanne was similarly constrained.

Cézanne, it is known, desired to paint the nude from life but was embarrassed by the female model—a fear of his own impulses which, when allowed free play in paintings from imagination, had resulted at an earlier time in pictures of violent passion. Renoir recalled to his own son years afterwards a conversation with Cézanne in which the latter had said: "I paint still-lifes. Women models frighten me. The sluts are always watching to catch you off your guard. You've got to be on the defensive all the time and the motif vanishes." Later when he carried out a large composition of women bathers in postures remembered for the most part from the art schools and the museums, he imposed on the faceless nudes a marked constraining order. . . .

By connecting his favored theme with the golden apple of myth he gave it a grander sense and alluded also to that dream of sexual fulfillment which Freud and others of his time too readily supposed was a general goal of the artist's sublimating effort.

Maybe you, like me, were astonished, in the middle of Meyer Schapiro's complex prose with its multisyllabic words, to find the word "sluts." But in my case, it evoked a vivid memory that suggested I should not have been shocked.

I was a student of Meyer Schapiro's. At Columbia College, in the spring of 1966, I was sitting in the auditorium where Profes sor Schapiro's wanderings had led him to Northern Renaissance paintings. He was showing us, projected from an old-fashioned glass slide onto the large screen in front, Jan van Eyck's *Rolin Madonna*, painted in about 1435, one of those marvels at the Louvre that never stops yielding miraculous pleasure and knowledge.

Our distinguished professor had carried on and on, breathlessly, and not just about the Madonna and Christ figures and the presentation of Chancellor Rolin on their same scale, making him appear to be a member of the Holy Family. He was also discussing—in detail and in language that I think no one could possibly follow, with adverbs like "theosophically" and "iconographically" and "hermeneutically" and "interrelatedly"—the highly detailed presentation of the Old World (the earth before Christ was born) and the New in the background of the painting. After explaining that the seven pillars supporting a bridge linking those worlds represent the seven sacraments, he called our attention to two figures standing with their backs to us in a courtyard between that distant bridge and the Madonna in the foreground of this complex masterpiece with its phenomenal rendition in minute detail. Suddenly, after what felt like ten minutes in which the dynamic speaker never caught his breath, he asked us what we thought one of those two tiny figures was doing as he faced into the heavens that separated the Old and the New. None of us dared answer. We assumed that the symbolism of the short man's way of looking, and the reasons we saw him only from behind, in his bright red cloak, were way beyond our understanding, even if we had memorized Erwin Panofsky and Jakob Rosenberg and the other scholars we had been assigned to read on the subject.

Professor Schapiro, hearing no answer, pulled on his right ear, as if to induce one. Still, no one, of the hundreds of people in the audience, uttered a word. Like a clap of thunder, at a high decibel level yet with a suddenly squeaky voice, as high as it was loud, with his flying white hair in the light of the slide projector and therefore magnified a hundredfold as filaments gone haywire over the exquisitely painted van Eyck, Professor Schapiro shouted out, "He's pissing! He's pissing!"

5.

In a footnote to this essay, Schapiro also reminisces about discussing apples with the great artist Alberto Giacometti, who often made this fruit his subject:

> I judge from conversation of Alberto Giacometti, who painted apples all his life, that he regarded the choice of the still-life object as an essential value and not as a "pretext" of form. In certain of his later paintings the apple has the air of a personal manifesto or demonstration—perhaps inspired by understanding of Cézanne's concern—as if he wished to assert dramatically, against the current indifference to the meaning of the still-life object, his own profound interest in the apple's solitary presence as a type of being.

Indeed, that is what the apple logo on the iPhone is—"a type of being"—just as the iPhone itself is "a type of being." And this is what the Bauhaus sought: objects that become true presences in our lives.

And the "type of being" is one that lives fully. Its missing bite, Adam having fallen for irresistible temptation even though damnation lies ahead, enters us into the territory of incomparable pleasures.

The embrace of earthly delights is worth the descent from paradise. At the Bauhaus, it was not even by choice. It was the Third Reich that destroyed a form of utopia. For Steve Jobs, it would be illness and too early a death. So why not bite the apple?

Part XII

1.

In 1997, Apple was floundering, with a serious risk of bankruptcy. Fighting to save the fledgling company, Jobs did the usual belt-tightening required at such a moment, which meant letting go of some employees and reducing expenses wherever he could. Yet at the same time he recognized the need for a public-relations campaign that might improve awareness of Apple's particular values as its founder saw them.

Right there, you have another salient point in common with the Bauhaus. The iPhone and the Bauhaus are today incredibly well-known, and their echoes are reverberating all over the world, more so all the time. Yet both had moments when they were hanging on to life by a shoestring. Gropius was forever scrambling just to survive. He had not wanted to do the major Bauhaus exhibition in Weimar in 1923, but the state authorities insisted that if the school was to continue to receive funding, he had to show what had already been achieved there, even though he thought they were not ready. Faculty and their spouses had to mop the floors of the exhibition space; no one had an extra cent. Steve Jobs was desperate for Apple simply to survive when he again turned to Ken Segall, his public-relations guru, and demanded an image that would distinguish Apple from other companies.

Segall and his team decided to emphasize Apple's radical-ism. Rather than follow the usual rules, the company developed unprecedented approaches to human needs. Apple needed to be known for its pioneering spirit. Craig Tanimoto, the art director at Segall's firm, came up with the slogan "Think Different."

The jolting bad grammar seems a deliberate provocation. The use of an adjective when an adverb is required has now become the norm, but in the late 1990s people were not yet saying "They did good" when meaning that a sports team won a match, or the now prevalent "I'm good" when they were telling waitpersons they did not want their coffee topped up. "Think Different" was an example of what it advised, jarring in form as well as concept.

That, of course, was the heart of the spirit of the Bauhaus. In the 1920s, few people, rich or poor, could imagine keeping their homes lean and mean. The goal of newly married couples—in an era when marriage meant the joining of a woman and a man, with no same-sex possibilities—was to make their homes *"gemütlich."* Lace curtains, heavy furniture carved with ornament, and flowery brocades were all there was. To opt for unadorned surfaces, and design based entirely on function, was the essence of thinking differently. Tanimoto would explain of the "Think Different" campaign: "It wasn't what we were looking for, but it was everything that we needed." That is a quintessentially Bauhaus attitude.

Rather than give himself credit for any special ingenuity, Tanimoto acted as if the idea for the game-changing slogan simply came to him. It was the same for the truest Bauhauslers. They never treated themselves as creators of ideas so much as the recipients of them. Klee and Kandinsky were grateful for inspiration without knowing its source. They were the opposite of the self-congratulatory "This is what I have done! Bravo me!" artists and architects of today. You could think that Mies's Barcelona Pavilion simply found its way onto the architect's drafting table. Anni Albers once wrote Florence Knoll, founder of the textile and design company Knoll, about a recent textile she wanted her to see: "Something new appeared on the loom."

From a Japanese family settled in America, Tanimoto was raised with a quintessentially Zen approach. He developed "Think Different" through quiet acceptance of what came his way. The timing was propitious. IBM was dominating the industry of personal computers, and Hitachi, Epson, Dell, and other companies were sowing their oats. This was that moment in 1997 when Steve Jobs, following twelve years away from Apple, had agreed to return. Cash was running out—fast. Tanimoto, whose family had been in internment camps during the Second World War, had been bright and tenacious enough to do well as a communications student at UCLA. Watching the Super Bowl in 1984, he was blown away by a commercial for Apple. Directed by Ridley Scott, who had made *Blade Runner* two years before, the TV ad shows a leader who resembles Big Brother. On a screen the size of a wall, this dictatorial ruler holds forth to an auditorium full of lifeless drones. A woman who is more animated than the rest charges forward and throws a sledgehammer at the screen, which bursts into a colorful fire. While Big Brother goes up in flames, an announcer declares that in this year of 1984, Apple guarantees that there will be no resemblance to the tyranny predicted in the novel George Orwell wrote in 1949.

Tanimoto was one of those rare people who are so inspired by what they see that they take the next step in life. Anni and Josef

Albers often said that there are many geniuses in the world, but few with the discipline and tenacity to make it matter. Tanimoto had the motivation and willpower they prized. He learned that the ad agency that had made this "1984 is not *1984*" booster for Apple was Chiat\Day. He approached the firm and persuaded them to hire him as an intern. But he had no intention of simply performing the tasks assigned to him. Tanimoto decided that while Apple had a brilliant logo, the company also needed a slogan. He mused. What did great innovators in various domains have in common?

Tanimoto's heroes included Mahatma Gandhi, Thomas Edison, and Albert Einstein. He realized they shared a brazen independence and courage. He needed to evoke that in as few words as possible.

Tanimoto would explain that "'Think different' celebrated individuality. People who marched to their own drummer, people who weren't afraid to be their own person. People who broke new ground." "Think different," like the name "Bauhaus," was a lean and sharp message to be unafraid and original.

2.

By the time Tanimoto developed his concept, it was well-known in the burgeoning field of communications that Thomas John Watson Sr., the genius behind IBM, had a sign saying "THINK" hanging prominently in his office. Tanimoto's slogan called more for inventiveness than for reflection.

Watson's background and education made Steve Jobs's seem privileged. The magic of "THINK" as what you faced in large bold letters when you met with him was the idea that brainpower alone could create machinery to change the world and make a pauper a millionaire. Born in 1874, Watson was one of five children whose father owned a modest lumberyard. Among his many jobs as a young man, he was a traveling salesman of Singer sewing machines. After a particularly successful day, he celebrated at "a roadside saloon"—the anecdote is told by his son, Thomas Watson Jr.—and drank until closing time. He emerged from the bar to discover that "his entire rig—horse, buggy, samples—had been stolen."

Watson went on to buy a butcher shop. Its NCR cash register interested him more than cutting meat did. He soon got himself a job with the maker of the device he admired, and began to climb the corporate ladder. All went well until 1912, when Watson was among the twenty-six NCR executives convicted for violating the Sherman Antitrust Act. He was sentenced to a year in prison.

Even a pardon from President Woodrow Wilson did not get him off the hook, but in 1915, the conviction was overturned and Watson's honor restored. He went on to become chairman of NCR, which was the first place where he made "THINK" the official company slogan.

Apple's "Think Different" was totally earnest; NCR's, and then IBM's, "THINK" had a certain irony. Whatever its intention, was Watson implying that a bit more thinking would have kept him out of jail for an antitrust conviction? There was an even worse disgrace in Watson's history where greater reflection would have been in order. In the same period when Herbert Bayer was designing Nazi propaganda, "IBM placed its technology at the disposal of Hitler's program of Jewish destruction." In 1937, Thomas Watson accepted from Adolf Hitler "the Merit Cross of the German Eagle with Star." It was said to honor Watson's punch-card technology, used for

In 1997, Apple Computer faced possible bankruptcy. Steve Jobs, who had been thrown out of the company twelve years earlier, agreed to return. In 1984, Craig Tanimoto, then a communications student, had been watching the Super Bowl when an Apple commercial directed by Ridley Scott, based on George Orwell's predictions in his 1949 novel *1984* of what life in that year would entail, blew him away. Tanimoto, whose Japanese-American family had been in an internment camp, was another individual who was generally undaunted, and who persevered tirelessly. He got himself a job with Apple's design department. Tanimoto's heroes included Mahatma Gandhi, Thomas Edison, and Albert Einstein, and at that moment in the late 1990s, when Apple was running out of cash fast, and Jobs was back at the helm, he came up with a slogan that linked all inventive and original people. The deliberate grammatical error—the word should have been the adverb "differently"—exemplified the message.

the 1933 census that identified Jews, and his continued idea that open trade between Germany and the United States could prevent war.

Watson soon shifted course. Following Kristallnacht, he wrote Hitler advising him "to give consideration to the Golden Rule in dealing with these minorities." In 1940, after Germany seized the Netherlands and started to invade France, he sent back his medal, saying it represented values antithetical to his beliefs.

IBM subsequently did a lot for the Allies throughout the war. But punch-card technology continued to help facilitate the horrors at Auschwitz. Also referred to as "Hollerith cards" after Herman Hollerith, who started IBM, this technology stored information in rows and columns of holes on cards that could be read by a tabulating machine to track people in a census that identified Jews. IBM published two special IBM punch cards for the SS and a third for the Nazi statistician who reported to Reichsführer Heinrich Himmler and provided Adolf Eichmann with information that aided the process of extermination. Who knows what "THINK" meant to Thomas Watson personally, or to some of the more savvy visitors to his office?

What Tanimoto proposed for Apple did not, of course, qualify the nature of what the approach might inspire. But neither that nor the grammatical issue of "Think Different" hampered its use. Lee Clow, the creative director internationally of Chiat\Day, immediately proposed reconsidering the incorrect use of "different." Tanimoto's response was even shorter than the slogan: "No." The two-word battle cry was subsequently submitted to Steve Jobs as the basis of an advertising campaign; Jobs went for it, and Chiat\Day produced it. Many consider this to have been the salvation of Apple as a company.

3.

The first iPhone hit the marketplace in the summer of 2007. By the end of the year, nearly four million had sold. This hand-held instrument has, as of this writing, gone through ten incarnations. Each is theoretically an improvement over the previous, but they are basically true to the initial design. The variations are minor, the color choices limited; the sense of identity is constant. There are infinite possibilities for cases—ranging from cheap and gaudy plastic ones with cartoon characters on them to crocodile- or buffalo-hide ones costing a fortune—but consistency rules.

These versatile but ostensibly simple objects, which embody the Bauhaus dream of equipping people across the earth with things that are well designed, clear, effective, and distinctly modern, have proved to be both a godsend and the curse of humanity. There are few greater vulgarities than someone reading a eulogy off of an iPhone; even if the tool that fits in a pocket is easier to manage than sheets of paper, it rankles when a speaker begins a homily to someone beloved, in a casket behind him, with "If I can get my phone to open . . ." People are shattered when a phone is broken or lost. Loving couples have vacations destroyed because of Internet failure that renders their iPhones impotent.

iPhones are not the Elgin Marbles, or Stonehenge, or anything else where part of the miracle is that it *lasts*. Apple's products have built-in obsolescence. Sooner or later, you have to face the need for an updated one. Yet, for all that is wrong, iPhones have established new possibilities of human conduct.

Of course, today they are only one of many devices that perform these functions; but the simplicity of their compact form is what makes them, more than other pocket-size models, fulfill much that was essential to the Bauhaus. Gropius ostensibly wanted to develop new buildings, but in fact there was never an architecture workshop at the school. The felicitous design of small objects was the true quarry of the school in both Weimar and Dessau.

The iPhone is evidence that visual charm, a tactile and optical harmoniousness, entertainment value, apparent minimalism, friendly materiality, and a correct sense of scale—so that the object in question is neither large enough to intimidate nor too small to be usable—fulfill human desire as well as need. This goal of the greatest design school of the last century was a tough bill of goods to realize.

The mentality is totally different from what happens, as is often the case in the modern world, when celebrity and glitz and conspicuous consumption captivate not only the few people who have them, but also the masses who have their faces pressed against the window of the candy shop of that wealthy minority. The products of Louis Vuitton, Ralph Lauren, Gucci, and other high-end designers are copied in cheap versions that are in turn sold to people of little means not because they are beautiful but because they are associated with financial success. The genuine handbags and sunglasses and other comestibles made by these fancy companies are affordable to few individuals, but hordes of people, especially teenagers, want to give the impression that they have the money to buy these objects that, above all else, flaunt their names and thus give status to their owners. The panoply of consumers has iPhones or the lower-priced clones not because of what they represent but because of what they do.

4.

With 2019 having been the Bauhaus centenary, its superficial image has proliferated. But now more than ever its truest values need protection. What was glorious in Anni and Josef Albers's house, with its inexpensive office furniture and thrift-shop cookies (still perfectly fresh, even if just past their "sell by" date), is seriously under threat.

An article about Jony Ive in London's *Financial Times* on October 19, 2018, written euphorically for its privileged readership, exemplifies the crisis. For those who see austerity and affordability and lack of chic as vital elements of the forms of beauty that last, the betrayal of those values is here. What has happened to the Spartan, even the puritanical? Where are the intelligence, and the refinement, of Shaker design? These were the real Bauhaus values. Why has infatuation with expensive luxury become so dominant? And the self-infatuation of the purveyors and consumers of what is rarified, yet often shoddy, ascended?

The *Financial Times* piece describes a lunch with Jony Ive at one of London's most select high-end restaurants, the River Café. The location itself is a bad start for anyone who cares about the essence of the Bauhaus as opposed to a false patina of its style. In honor of its own thirtieth birthday, this glossy eatery purportedly used Josef Albers's stencil lettering for outdoor signage and throughout a cookbook. One might have assumed that this was simply a borrowing of Albers's beautiful alphabet, of which the original glass letters are at the Museum of Modern Art in New York, and which have been reproduced and exhibited in numerous places as the Bauhaus masterpiece they are. The River Café, however, did not simply adapt an idea; they claim, in the cookbook, that this is indeed Josef's lettering. The main offense is not that they failed to obtain the requisite permission to do this. The travesty is that they changed the proportions of each of the letters so as to render them illegible. Josef's alphabet is a model of careful dimensions based on perfect squares and circles, and parts or combinations thereof. Josef kept these underlying units meticulous and exact, and organized them so that the resultant letters are perfectly readable. The River Café has taken this alphabet and compressed and elongated or widened and fattened the forms, so that what should be perfect half circles are now like half circles in a funhouse mirror. They have changed the dimensions of the openings between elements, widening them so that the end results are almost impossible to read.

To be true to the Bauhaus is not to follow a "style." It is to maintain impeccable standards, consider every nuance, and make successful functioning the priority. In 1974, Josef Albers told me that he had never been satisfied with the *z* of the alphabet he made in 1930. He had tried a mirror image of the *s*, but it did not read clearly. He had then worked and reworked triangles (that were precise half squares) and semicircles, but still the *z* was not sufficiently legible. Saying this to me, Josef handed over a photograph of his alphabet—the version in milk glass owned by the Museum of Modern Art. He had penciled in a *z*—there is none in MoMA's version—following the *y*. The pencil on a glossy photo was faint but clear. "At last, Nick! I have only now figured out the *z*! And you are the keeper of the *z*."

The interview with Jony Ive took place at "an outside table" at the restaurant where reservations have to be made well in advance. Did Jony Ive notice the disturbingly disproportionate signage overlooking the Thames? Does he still have the refinement of eye he had at his modest start?

If so, we do not learn a word about it in the profusion of details about the lunch, whose menu and prices per item are provided: "Total (inc service and donation) £183.69." This is not pocket change, given that there was no wine consumed, but nota bene: "Natia mineral water" and "Ferrarelle mineral water," each at three pounds fifty a bottle.

In 1923, when they had both arrived in the previous year or two at the Weimar Bauhaus, Paul Klee and Wassily Kandinsky went out to a café together. This was the epoch of such sharp inflation in Germany that people carted their Deutschemarks in wheelbarrows. The two great painters, old friends since Munich days, were about to order, and then they realized that they could not afford the cost of two cups of coffee. There was no thought of bread or pastry: they simply wanted coffee. Knowing they did not have the necessary money, they left and went to Kandinsky's apartment and had their coffee there, from the samovar that was one of the few remnants of the world from which he and his young wife, Nina, had come.

Nicholas Foulkes, the writer lunching with Ive at the River Café, enumerates the astounding capacities of the Apple Watch, and continues:

It sets me wondering what Apple consumers would make of the designer's wardrobe as he makes his way along the outside tables at an amiable amble. The 51-year-old is wearing a suit tailored by Caraceni of Milan in a lightweight *pied-de-poule*, a white linen shirt and his signature Clarks Wallabees. He over-apologises for being 10 minutes late. He spots the architect Lord Rogers at the next table; there is an outbreak of mildly abashed mutual effusiveness; then he settles into his chair, picks up the menu and lets out a sigh of satisfaction.

The italics of "*pied-de-poule*" are, we assume, because the term is French. But who among the readers, including French speakers, knows what these words, which translate to "chicken's foot," mean in a suit fabric? And even those who can name a number of good Milanese tailors are not necessarily familiar with Caraceni, although it clearly has to be one of the best. Are the Wallabees a reference to Ive's humble origins in northern England, a way of saying that, even if he now wears *pied-de-poule*, he is an honest lad who walks at a fast clip?

What is this nonsense? It is interesting to learn that the designer of the iPhone is now chancellor of the Royal College of Art, but how much name-dropping do we need? We learn that Ive initially met Ruth Rogers, who owns the restaurant, at the Obama White House. We are told what the clothing magnate Paul Smith said about "Ruthie"—wife of the architect previously identified by his title—and that Ive attended Princess Eugenie's wedding.

One could, by the way, do an amazing "connect-the-dots" between all these famous folk and Anni and Josef Albers. Michelle and Barack Obama selected two paintings by Josef to borrow for the White House shortly after he became president, and the Albers Foundation subsequently gave two others and a large rug by Anni to the permanent collection of the presidential residence, specifically in honor of the Obamas, who were as excited in return as one could hope. Paul Smith is a longtime Bauhaus enthusiast, having discovered the work of the school at a Bauhaus exhibition in London in 1969—he still has the catalogue—and collaborated with the Albers Foundation on products connected with Anni's work. Princess Eugenie's large wedding may have been as traditionally English in style as you can get, but any number of people on the guest list work in fields with direct Bauhaus connection. Still, there is something all wrong when we are told that "it says much for the influence of Apple's designer-in-chief that he is now mixing with presidents and royalty."

And then we learn about Jony Ive's Rolls-Royce Silver Cloud and Aston Martin DB4. When Anni Albers was still Annelise Elsa Frieda Fleischmann, a young weaving student at the Bauhaus in Weimar, two of her uncles came to visit her in a Hispano-Suiza. Her immediate response was to signal them to get that fancy car out of sight of her fellow students and anyone else at the Bauhaus; she was desperate not to be seen as wealthy, and, worse yet, pretentious.

Anni was happy enough when she could greet her uncles more amicably on a side street. She willingly accepted their gift of "valuta"—hard currency—that enabled her to purchase a watercolor by Paul Klee, who thumbtacked his recent work to the corridors of the school every few months. But showing off was never the point—quite the opposite.

Functionalism, clean design for everyone, a sense of the universal, and higher values. As we consider the Bauhaus a hundred years after its creation, these are what counts.

5.

It was big news in the business pages in late June 2019 when Jony Ive announced that he was leaving Apple. The market value of Apple stock dropped in heavy trading, and analysis after analysis evaluated the reasons and the impact on the future. Names again are a big part of the story. In the English press, at least, Ive was now referred to as Sir Jonathan. The man who has been CEO of Apple since 2011, following Steve Jobs's death at age fifty-six, and whose operation of the company may or may not have been the reason for Ive's move, has, meanwhile, kept up the company's signature informality by being known to everyone simply as Tim Cook. (He is Timothy Donald Cook.) Ive has created a new firm that, everyone says, will work closely with Apple. Here the problems begin. Its name is LoveFrom. What is certain is that this is a long way from the restraint and dignity of Bauhaus.

Part XIII

1.

Several of the Bauhaus artists consciously took their painting
into the realm of the cosmic. Even more than any of the others,
Paul Klee inhabited that territory of infinite space and endless
time and evoked it in his paintings. Klee connected with *everything*. He was one with the birds chirping outside, with the fish
in the sea, with the distant stars. And the watercolors and oils
he produced in his years in Weimar and Dessau render whatever he painted—minarets in the distant desert, flower gardens
under the evening stars, complex imaginary cities, ships depart-
ing toward the unknown—sublime.

Anni Albers said that Klee walked around as if in another
world. Her preferred analogy was that he was like "Saint Chris
topher with the weight of the world on his shoulders"; she
invariably said the *"Chr"* of "Christopher" as if she were slightly
clearing her throat, a soft rolling *Chaaa* before the *"riss."* For
Anni, this suggested both his distance in another orbit and the
burden of being there.

It was slightly off as an analogy, but the gist pertains. The
legend is that Saint Christopher was a giant who converted to
Christianity and had the task of standing on the banks of a river
for which there was no bridge or boat making it possible to get

across. The enormous saint ferried travelers over the rushing water simply by hoisting them onto his broad shoulders and carrying them. His gigantic legs were long and sturdy enough to withstand the greatest depths and harshest current.

But strong as he was, one day when he had a small child up there with his legs dangling in front, the child who had been light at the start became so heavy that Christopher had to struggle with all his might to reach the distant shore. This was because the child was Christ himself, and the weight so crushing because it was that of all of the world's sins.

It's a myth with endless ramifications. Christ may have borne the burden of the world's sins—with Christopher (the named means "Christ bearer") assuming them in turn—but he had redeemed those sins by taking them on. What Anni Albers meant was that Klee was otherworldly, absorbed in his own sphere, not quite of the earth, and that he seemed beleaguered by his task. If there was a relationship to sin, Klee's connection with avarice or pride or envy or any other of the travesties of human attitude and behavior was to relieve them so that the culprits could be freer.

That freedom is the essence of going beyond oneself. To recognize oneself as a tiny element in the vast cosmos is liberation.

In 1929, when Klee was approaching his fiftieth birthday, Anni heard that three other students in the weaving workshop were hiring a small plane from the Junkers aircraft plant, not far away, so that they could have this mystical, otherworldly man's birthday presents descend to him from above. He was, they had decided, beyond having gifts arrive on the earthly level where ordinary mortals live.

Klee's presents were to come down in a large package shaped like an angel. Anni fashioned its curled hair out of shimmering brass shavings. Other Bauhauslers made the gifts the angel would carry: a print from Lyonel Feininger, a lamp from Marianne Brandt, some small objects from the wood workshop.

Anni was not scheduled to be in the four-seater Junkers aircraft from which the angel would descend, but when she arrived at the airfield to see the other weavers take off, the pilot deemed her so light that he invited her on board. For all four passengers, it was their first flight. As the cold December air penetrated her coat, the pilot fooled with the young weavers by doing 360-degree turnabouts. As they huddled together in the open cockpit, Anni became unexpectedly aware of the new visual dimensions. She had been living on one optical plane in her textiles and abstract gouaches. Now she was seeing from a very different vantage point because of the ever shifting perspectives of familiar buildings and parks of Dessau as she viewed them from above. She was too fascinated to be afraid.

Anni Albers guided the pilot by identifying the house the Klees shared with the Kandinskys, next door to where she and Josef lived. He swooped the plane down and they released the gift. The angel's parachute didn't open fully, and it landed with a bit of a crash, but Klee was pleased nonetheless. He would memorialize the unusual presents and their delivery in a painting that shows a cornucopia of gifts on the ground in good condition, even if the angel looks a bit the worse for wear. James Thrall Soby, a pioneering art collector and curator and writer who owned the colorful canvas, told my wife and me that he thought it depicted "a woman passed out drunk at a cocktail party," with the golden brushwork Klee used for Anni's brass shavings representing the socialite's blond hair, but Anni's account revealed the actual facts.

Josef Albers was less impressed than Klee was. Later that afternoon he asked Anni if she had seen "the idiots flying around overhead." Anni smiled mischievously as she recalled this. "I told him I was one of them," she said with her usual tone of proud defiance.

This was the spirit of the Bauhaus. Yes, thinking differently. Using industry as a pathway to playfulness. Trying what no one else had considered.

2.

The world beyond our individual earthly existence was also in Josef Albers's thoughts when he made a blue-and-green twenty-four-inch *Homage to the Square* in January 1976, some two months before his death.

By the time he created this *Homage,* the eighty-seven-year-old Albers was working on very few paintings. His hand was too unsteady; printmaking, executed with machinery, was easier. Still, he painted this panel as a study for an Aubusson tapestry that had been commissioned for a cavernous bank lobby in Sydney, Australia.

Albers had a problem making the painting. He had found a combination of his chosen colors that interacted perfectly in a format with the central square relatively small, but that did not work as well with the central square larger. In the version with the larger middle square, "downstairs" was fine, but "upstairs" was "hell." He wanted both a spatial flow and a color "intersection."

The intersection occurred in the version with the small central square, but Josef was not satisfied. Moving his hand over the sky-blue center, and then over the more terrestrial forest green and the oceanic aqua surrounding it, he explained that the middle color was the cosmos, surrounded by earth and sea. The problem was that with the middle square small, the cosmos appeared too distant.

Albers wanted all boundaries and edges between the colors to disappear optically. Additionally, there should be no sharper corners on the inner square. (He said proudly that Cartier-Bresson once told him that he made "circular squares.") To achieve these effects he needed to find colors with the identical light intensity.

He experimented with the various cobalt greens he had in his trays brimming with tubes and paint in every hue and by almost every manufacturer who made oil pigments. He discovered that Winsor & Newton's Cobalt Green, no. 196, was perfect. But Winsor & Newton's current Cobalt Green, from batch no. 205, failed.

No supplier had the older paints. The head of Winsor & Newton's US distribution said there was no difference; that the newer and the older Cobalt Greens were totally consistent. But then he learned that Josef Albers perceived an essential disparity in the light intensity, and that only paint from the previous batch served his purpose. The paint executive found five tubes of no. 196 in stock and was thrilled to send them to Josef.

When Josef made the *Homage* with the precise pigment he wanted, he saw the corners and boundaries evaporate, and the adjacent colors penetrate one another. He explained that he had to have the format with the larger center because it made the cosmos more present. "For me, the cosmos is getting nearer," he said. He finished the painting in February 1976. It was his last work of art. He died the following month.

Know your materials! See the most minute nuances! Savor them! God is in the details.

This too is the spirit of the Bauhaus.

The sky's the limit, and also limitless. This is the thinking behind the iPhone at its best.

Josef Albers painted more than two thousand *Homages to the Square* in the twenty-six years after he developed the format in 1950, at age sixty-two. For the last one he made, which he finished in February of 1976, he struggled to find the right paint for the central blue. It represented the cosmos to him, and his problem was that Winsor & Newton's current Cobalt Green did not achieve his objective of having the boundaries and corners evaporate. The new Cobalt Green succeeded if the middle square was kept small, but Josef wanted it large because he felt that the cosmos was "getting nearer" to him. He finally managed to get five tubes from the earlier batch of Cobalt Green, which enabled the edges to soften and the adjacent colors to penetrate, now with the cosmos larger and seemingly nearer. Josef died a month after completing the masterpiece.

3.

Kandinsky's abstractions are also about the infinite forces and elements of the universe itself, the multiplicity and complexity of it all.

The relationship of the individual to totality is an obsession of just about every human culture. We all belong to what is boundless and eternal.

This is Plato's notion of anima mundi, the soul that pervades everything in the world. Jung and his disciples noted it in a piece of wood that might be used as a writing surface, in an armchair, or, for that matter, in anything whatsoever. Everything is part of something greater. And everything is connected, within the same vast web.

4.

When visitors arrived at Paul Klee's studio at the Dessau Bauhaus, he did not show them his latest paintings facing the wall. He had people look inside his aquarium rather than at his art; Klee had them observe the miraculous markings on tropical fish, and their amazing forms. He would gently tap on the glass so as to get one bright blue striped specimen to move along so as to reveal the orange parrot fish behind it. In the kitchen, one of his favorite haunts, he would invent recipes of stews made from innards, his steamed lung ragout one of his specialties. Everything could be used, everything had a function, everything worked. He painted fish that ate people, and cooked for people who ate fish.

When Klee and his wife, Lily, had guests come over, they hardly spoke. Rather than converse with his guests about matters of no lasting significance, Klee took out his violin, Lily sat at the piano, and they played Mozart duets. The music was timeless, its movement never-ending. One was transported emotionally in a way possible for all people everywhere.

Kandinsky took you to the stars as far away as the eye could see. At the same time, he enabled you to enter the human mind. The artist is a single individual, and he or she makes a single object at a time, but infinity awaits. Kandinsky said, "There is always an 'and.'"

This, too, is the world within the iPhone. Of course, not the iPhone solely—but the iPhone, more than any other instrument of the technological revolution, contains more potential, reaches farther, into vaster realms, per square inch of its tiny self, than anything that had been developed before. From the small device the size of a playing card, the thickness of perhaps a dozen such cards stacked on top of one another, one might telephone from Japan to Chicago, or access information on prehistoric art, or learn how amoebas metabolize the microorganisms on which they live. This adheres to the same underlying principle of all Bauhaus design and art. What is visually pleasing, agreeable, and enticing—and friendly and simple in appearance—is both a part of the infinite vastness of all and everything, and a link to it.

Bauhaus designs at their best, and the iPhone, work to fulfill the Platonic ideal of form. They do not all achieve the refined, understated perfection that was the Greek philosopher's sine qua non, but each in its own way is sublime. Mies's Villa Tugendhat is composed of large expanses of glass and ebony in glorious juxtaposition. Anni Albers's rhythmic wall hangings cause stripes and rectangles to vibrate joyously. Bayer's Universal lettering style is lean and graceful. Josef Albers's sandblasted glass constructions consist of flat blacks and whites that syncopate at right angles to one another musically. Paul Klee's colorful mosaics of slightly irregular squares look as if they have been conjured by a magician. Kandinsky's triangles and lines and circles gyrate so as to suggest human complexity made orderly. Brandt's glass spheres and tensile stainless-steel armatures are immaculate. Breuer's kitchen cabinetry makes the inanimate

Since its launch in 2007, the iPhone has brought out a series of new models. To date, it keeps getting slightly larger—like a child growing up (in this case slowly). Like a true classic, for all its advances within, on the surface it remains essentially the same—in its sleek sheath, with the overall covering remaining simply a new variation of itself. This is true to the Bauhaus mentality: if a form has been carefully developed, and succeeds, one should stick to it. In painting and printmaking, Josef Albers's *Homages to the Square* exemplified this idea of a simple, premeditated shell remaining the basis of a series.

dynamic. Oskar Schlemmer's conical costumes and sets for *The Triadic Ballet* sparkle and enchant. A sleek iPhone with its perfect facade houses complex networks within networks inside. These carefully conceived visions encapsulate the ideals elucidated by Plato.

5.

Plato in a discussion of the iPhone as well as the Bauhaus?

Educated people know the name of the Greek philosopher, but few realize how much his observations and standards pertain to modern civilization. Plato's values pervade especially the best of the Bauhaus and the best of more recent technological instruments.

Start with Plato's "The wisest have the most authority." At Aspen, Steve Jobs spoke specifically about the five thousand geniuses who would effect change for the world at large. At the Bauhaus, the masters were like the Pantheon of top-ranking thinkers. There was no question about a hierarchy of intelligence. The ability to see the present clearly and forecast the future wisely gave "the wisest . . . the most authority."

One cannot say that this has actually been the case throughout human history. We have had Hitler and Donald Trump and others who, while "wise" in their ability to deceive, have had the form of knowledge that rendered their authority catastrophic. In the development of what is best in modern design, however, the wise elite that have succeeded have been well intentioned as well as brilliant.

Plato's notion of beauty extends beyond the design of objects. Rather, it reflects the abiding faith, the love of the good, and the idea that visual beauty is in accord with spiritual beauty, which is so essential to our story: "For he who would proceed aright . . . should begin in youth to visit beautiful forms . . . out of that he should create fair thoughts; and soon he will of himself perceive that the beauty of one form is akin to the beauty of another, and that beauty in every form is one and the same."

Then there is this simple aphorism of Plato's, to some so basic that it falls into the category of the merely obvious; yet when we think of the Bauhaus, and of the iPhone, and of the importance of a clear initial concept to all that follows, it is on the mark: "The beginning is the most important part of the work."

On the issue of construction, one of Plato's more provocative declarations has particular validity: "As the builders say, the larger stones do not lie well without the lesser." This is to say that every single element—the corners that Jonathan Ive reconfigured until he got them right for the iPhone; the rivets that Marcel Breuer chose for his armchair; the small dashes of paint that animate the celestial landscapes of Paul Klee—has been accorded total focus, and that the finished object can be seen as the compendium of the small components that are its building blocks.

Then there is Plato on human happiness. "The greatest wealth is to live content with little."

At the Bauhaus, Josef Albers wrote a friend for whose apartment he was designing furniture, "An empty room is always the best." In Connecticut, Albers's bedroom had a single bed, a bookshelf that served as a bedside table, a desk, and a desk chair. He needed nothing more. The only thing on the walls was a small thermostat.

The thermostat itself was in a gold plastic circular case that was patently unattractive. Its capacity to regulate the heat and air conditioning that were blown through discreet vents was, however, more important than its aesthetic limitations. The iPhone is intended to be an object that, if it is your sole possession in rural Africa, is such a major enhancement to your life that you ostensibly need little more. Of course it is understandable that most people desire a great deal more, but the contentment afforded by the iPhone compensates to some extent for the lacunae.

6.

Sometimes, it is some of Plato's simplest aphorisms that apply. They link the modern technology of which the iPhone is the summit with the goals that motivated the Bauhaus workshops. Consider the straightforward "There is no harm in repeating a good thing." Textiles produced by the yard in the Dessau weaving workshop, Breuer's side chairs that made it into production, the enticing postcards that Klee and Kandinsky made and printed for the 1923 Bauhaus exhibition: as with iPhones coming off the assembly line by the million, modification or individuation would only have hampered the spreading of what was totally worthy as it was, to be enjoyed by large numbers of people.

Or Plato's "Courage is a kind of salvation." It is, of course, the rallying cry of many: in poetry, W. E. Henley's "Invictus," Robert Browning's "Andrea del Sarto," and Robert Frost's "The Road Not Taken"; in national leadership, the intonations of Winston Churchill and John F. Kennedy. Still, courage remains shockingly rare. The world is full of followers, of people without the will or the desire to go against the tide, of keepers-up with the Joneses. It took sheer courage for Anni Albers to buck the tide by going to the Bauhaus and take such a different course from all the other women in her family, who gladly accepted lives of running lavish households with ample staff. To opt for a single rented room with the chance for a bath only once a week, as Anni did, required steely nerve. It required bravery in abundance for Walter Gropius not only to have survived the aerial bombardment that killed his troop mates on every side of him but, almost immediately afterward, to start a school that endorsed standards contradictory to all that had preceded it. It also took courage to drop out of Reed College and keep making computer models in the family's garage, and fierce originality to insist that good design was a more vital goal than profit margins. Few people have that extreme courage that Plato recognized as "salvation."

Relatedly, there is great perspicacity in Plato's "Courage is knowing what not to fear." The most tenacious souls of the Bauhaus and the technology world did not fear rejection, mockery, or even potential failure. The courage strong enough to overpower these fears shines in the resultant objects.

There are more surprising views codified by Plato that also pertain to what Bauhaus ideals and iPhones share. "Everything that deceives may be said to enchant" is a case in point. Loathsome as deceptive human behavior can be, the idea of things that deceive is something else. At the Bauhaus, it was Josef Albers who, more than any other artist, relished deception—in color and line. The way that the same yellow appears grayish against a brighter yellow and more truly yellow against a gray was his lodestone. The way that the identical color can present two very different appearances was a miracle worth revealing in various forms. This is again because of that essential factor of "perception" Jonathan Ive deemed fundamental to product design.

Albers made constructions where straight lines appeared curved. He drew lines that at one moment appear to tip backwards away from us from right to left and then appear to move conversely, going away from us from left to right: a physical impossibility. Deception in the world of computers occurs constantly, in how the apparently immaterial and nonexistent summons itself to life with a few movements of the fingertips, or in the way that a person who is thousands of miles away seems to be in the same room with FaceTime.

Plato advocated discipline and perfectionism, the effort to produce quality in small quantity rather than a vast range of what is second-rate. His "Better a little which is well done, than a great deal imperfectly" pertains to the stage after stage with which the artisans in Bauhaus workshops would try and try and try again, adding and removing components until getting the ideal result. It underscores the development of any of the Apple products that finally make it to market, with the iPhone the exemplar. Using what is sloppy or slapdash is a betrayal of creator and consumer. Honest effort and intelligent refinement enter our very beings as we behold or utilize objects created with them.

7.

Platonic thinking is ingrained in all of us. Far-ranging, incalculably brilliant, it also has great familiarity. Plato encapsulated complexity in simple form, made diligence look nearly effortless, and gave faith in the quality of human existence.

We have scant information about the man himself. His birth and death dates are uncertain. He was born roughly at the start of the last quarter of the fifth century BC and died near the middle of the fourth century BC. He founded the Academy of Athens. He developed and helped spread an approach to human life that gave classical Greece the luster and aura of wisdom that have made it a reference point for greatness ever since.

Plato shared, with his teacher Socrates and his student Aristotle, an appreciation of human life and a wish for great numbers of people to enjoy it selflessly and intelligently. In his *Republic* and *Laws* and other writings, Plato articulated what was once called the "European philosophical tradition." Now, with increasing awareness of the universality of much human reflection, the coherence of philosophy and spiritual beliefs from every corner of the earth, "Western" society is beginning to overcome its arrogance in treating itself as the origin of what is inherent in all human cultures. The earth has become significantly more united between the time of Plato's writing and the creation of the Bauhaus, and even more so since the closing of the Bauhaus in 1933. The technological revolution that began in the aftermath of World War II has given new meaning to the word "global." "Global" is the operative term with the iPhone, which is designed for everyone, everywhere.

It was Plato who said, "Necessity is the mother of invention." This simple idea, even though a cliché, is totally valid. It applies to Bauhaus tables as to the iPhone. While none of these accouterments to life are as necessary as oxygen and rudimentary nutrition, they fulfill essential human needs, whether physical or emotional.

Plato's preferred manner of discourse took the form of dialogues. Perhaps the shortest, and among the richest, is this one: "And what, Socrates, is the food of the soul? Surely, I said, knowledge is the food of the soul." No one who persevered at the Bauhaus or at the research department of Apple Computer would have argued. What is vital is the principle of Bauhaus pedagogy and the education received by people like Steve Jobs and Jony Ive and Craig Tanimoto: that knowledge extends far beyond the accumulation of facts. It is both more profound and more entertaining. Plato also said, "Life must be lived as play." The toys constructed at the Bauhaus, Anni Albers's approach to weaving, the puppets Paul Klee made for his son, all embraced that playfulness.

If necessity is the mother of invention, then having a good time must be recognized as necessity. The joy of engaging in actions that result in sheer pleasure, without purpose, is vital. It adds to the rich tapestry of life.

Klee's son, Felix, used those puppets in performances for audiences of Bauhauslers in Dessau. The adolescent made those colorful creatures he animated with his hands into personalities everyone knew. Felix had heard and understood lots of gossip his parents exchanged with the Kandinskys. The Bauhaus "grown-ups" in the audience recognized themselves in husbands and wives squawking at one another and in fashion-conscious characters snidely critiquing the new clothing of their rivals. Felix also orchestrated imitations of masters giving lectures, the teachers shouting bombastically.

If in its range of games the iPhone encourages less inventive forms of playfulness, it fulfills the same need for lack of purpose and for recreation.

Fun is esteemed in life. This has not always been, and still is not, the governing approach in every culture. It is, however, the attitude shared by the Bauhaus and the doyens of Silicon Valley.

Plato was serious about happiness; he emphasized the emotions of human pleasure. "Love is the joy of the good, the wonder of the wise, the amazement of the gods." And while the Greek believed in those gods who, in our own time, constitute what we call "mythology" and treat as the stuff of fiction, he wanted that amazement to be everyone's. "If particulars are to have meaning, there must be universals." One of those universals is the need for the positive in everyday life. In the era of expressionism and the New British Art and industry devoted to the weapons of war, that focus on what can be celebrated is not a given. It is one of the links of the Bauhaus spirit and the best of the achievement of products like the iPhone made to enhance daily living.

8.

Plato relished imagination. "The man who arrives at the doors of artistic creation with none of the madness of the Muses would be convinced that technical ability alone was enough to make an artist. . . . What that man creates by means of reason will pale before the art of inspired beings."

The creative spark is both inexplicable and glorious. The capacity to go beyond the sphere of most people's thinking, to leap from nowhere to an unprecedented vision, defies understanding. But even as the origins of Plato's "madness of the Muses"—of artistic imagination—elude us, that spirit is manifest in its by-products. Every time we sit in a Barcelona armchair and feel supported by its flawless planes of leather positioned precisely to accommodate our bodies, with the chrome X's making the structure of the whole like ballet dancers in a pirouette, we imbibe Mies's wonderful willingness to brave the unknown and cultivate its charms. Genius and originality become part of us whenever we pick up that crazy little iPhone.

Plato, equally, gave due honor to control. Flights of fancy only succeed when they are realized with discipline and refinement. Steely know-how and paring down are as fundamental to Jony Ive's design as to Herbert Bayer's lettering and to Erik Dietman's furniture. "Excess generally causes reaction, and produces a change in the opposite direction, whether it be in the seasons, or in individuals, or in governments," Plato wrote. The deliberate confinement to the essential without the fluff that results in the most effective material objects has echoes in the best of all that is natural and human.

What we see and touch and use can either reinforce what is true morality or contradict it. An interchange occurs between the qualities of art in every form and all of human behavior. Excess, like irrelevancy in design, apparent in anything from a can opener to a painting on the walls of a museum, is a form of corruption. If the artists and designers at the Bauhaus were alive today, the sleazy facades of buildings like Trump hotels, the silliness of the latest products by Louis Vuitton with their vulgar patterns, would not merely be distasteful, but emblematic of distorted human values. Falseness in appearance is, to many of us, innately disturbing, and causes us to flinch or to react in annoyance. Fealty to functionality without an iota of waste achieves Platonic perfection.

9.

From Plato: "The measure of a man is what he does with power." No one would disagree that not just Steve Jobs and Jon Ive, but others in their world, fill the bill. This was the same tenacity that Anni and Josef Albers emphasized as central to success, and that all at the Bauhaus recognized as being essential if creative genius was to have meaning. And then there is Plato's splendid exhortation to hard work: "Apply yourself both now and in the next life. Without effort, you cannot be prosperous. Though the land be good, you cannot have an abundant crop without cultivation."

The Bauhauslers exulted in well-developed vineyards and fields of crops. For Steve Jobs, the apple orchard was more important to his inner formation than was any educational institution. Even for those of us for whom Plato's notion of "the next life" is alien, the emphasis on cultivation is halcyon.

10.

Those of Plato's aphorisms that initially puzzle you, and therefore force you to reread them, are often the ones that, in the

Oskar Schlemmer's 1932 *Bauhaus Stairway* takes us inside the school where women and men banded in a community to advance a common cause. Their individual selves, and faces, mattered less than the constant pursuit of new designs to serve all of humankind.

combined playfulness and perceptiveness that makes them so complex, bring the greatest insight. "To suffer the penalty of too much haste, which is too little speed": this stumps us at first. But, then, imagine life within the Bauhaus "workshops" in Weimar that in turn were named "laboratories" in Dessau. Imagine the efforts in the locked-up facilities where no unauthorized people could even enter the secret universe in which the iPhone was being developed in Cupertino. "Haste" would have been frantic, and borne no fruit. "Speed" was essential. Even if we do not know Greek, and cannot grasp the precise distinctions between the words Plato originally used, in whatever language this statement is translated, it is superb advice for all of us.

Plato did not invent his value system, but he articulated it as had no one before. "A man's duty is to find out where the truth is, or if he cannot, at least to take the best possible human doctrine and the hardest to disprove, and to ride on this like a raft over the waters of life." That drive and that integrity are essential to the protagonists of this book. They took their tasks to heart; they achieved their goals through constant dedication and uncompromising standards. No shortcuts or sleights of hand entered the picture. Plato further codified their genuine and ceaseless trying for the best when he wrote, "Cunning . . . is but the low mimic of wisdom." Even more than Socrates or Aristotle, Plato was the source of the philosophy manifest in the particular achievements celebrated in this book. "Wise men speak because they have something to say; fools because they have to say something." Of course the gender-specific "men" is unfortunate. But when you imagine the voice of Walter Gropius welcoming those Bauhaus students, when you read Anni Albers's texts on designing, and when you hear Steve Jobs's talk at the Aspen Institute in 1983, you hear people with something to say. The concision, pursuit of truth, and elimination of the gratuitous in their communication is as evident in what they produced as in what they said.

One of Plato's "absolutes" fails to apply either to the Bauhaus or to the American technology, however. The Greek declared, "I never did anything worth doing by accident, nor did any of my inventions come by accident; they came by work." The Bauhauslers, on the other hand, allowed for accident, and sometimes cultivated it. Anni Albers thrilled to items placed on the wrong shelves in department stores—a screwdriver among the cooking implements, for example. The unintended juxtaposition might lead her to the use of the screwdriver as a stirrer for iced coffee, its graceful proportions and yellow plastic handle and shiny metal blade better suited to the task than anything intentionally designed for it. Or maybe the screwdriver could be moved gently through wet ink on a lithographic plate—as a device for a new process that could be the basis of an abstract print.

She produced one of her most glorious prints following the sheer happenstance of a negative of her latest print falling by chance on top of a Velox of the same on her work table. The negative was about a half inch off the printed image underneath it. Anni loved the unintended interplay when the identical design, on transparent material, was on top of itself at both a horizontal and a vertical shift. The accident became the basis of a masterpiece.

When Corning made Gorilla Glass, and when the Apple team tried this or that rubber gasket around it, they were alert to what could occur without intentionality. When Le Corbusier saw a salt and a pepper shaker on the table of a small restaurant where he was having breakfast in the Ahmedabad airport, in the time period when he was trying but failing to come up with the right form for the large Punjab Congress Assembly Hall he was designing in Chandigarh, those plastic objects with which he flavored his fried eggs provided his solution.

But Plato, for all his love of the visual, was a philosopher, not a designer of buildings or objects or a creator of paintings. If accident had no value for him, that is less significant than the value that he places on "work." Work to the bone—not to have more money in the bank through financial manipulation, but to contribute to civilization. Without diligence and stoicism, Martin Luther King would not have transformed society. Jonas Salk would not have developed the polio vaccine. The current trends of taking it easy and the emphasis on avoiding "stress"—a degree of which is essential to achievement—are not part of this story.

11.

What is perhaps Plato's best-known broad-sweeping profundity poses more difficulties than many others. This is the Greek's famous "The good is the beautiful."

The question it prompts is whether Plato wants us to see that being good makes a person or object beautiful, or that appearing beautiful makes someone or something good.

Almost all that was produced at the Bauhaus—from flatware to lighting instruments to kitchen cabinets—is of such quality as to be "good." The iPhone, regardless of its irritating elements and its dubious impact on human relationships, is also so ingenious, and technologically and aesthetically impressive, as to be considered "good." All of these objects, made to be the best they can be, and conceived and executed with overarching intelligence, belong, therefore, to the aspects of existence that are "beautiful." And their beautiful appearance makes them good.

There is a lot of corruption and moral ugliness and selfishness in the world. This "good" as "beautiful" and "beautiful" as "good" is a sublime alternative.

Acknowledgments

My thanks go first to the dedicatees of this book. For more than fifty years, Charlie Kingsley—always "Charlie," although of course a "Charles," because, if born to formality, and invariably correct, he is quintessentially friendly—has been a stalwart of the joyous work first, of making Anni and Josef Albers's lives go more easily, and then of preserving their legacy. Gretchen and Charlie, like the Alberses, are another married couple united by an ethos. They exemplify compassion, generosity, integrity, independence, kindness, and moral bravery. From the time our adult daughters were little girls, and now that our first grandchildren have entered the scene with extraordinary panache, all of us Webers have had our lives enhanced immeasurably by the supportive, spectacular Charlie and the ineffably warm, sparkling Gretchen, an individual whose politics and commitment to beliefs reveal the empathy and sense of "the other" that she and Charlie have in spades.

As a septuagenarian, I have come to treasure, especially, those relationships that not only survive but flourish with time. Anni and Josef Albers, whom I met half a century ago, remain inspiring well past their deaths—and their art and shared beliefs promise to add luster to human existence indefinitely. Victoria Wilson remains the most perspicacious, tenacious, intrepid, and insight-

Acknowledgments

ful of editors. Vicky has the quality of Anni Albers of seeming, in Anni's words, "always the youngest one in the room," because of her perpetual originality, verve, and unwillingness to compromise. For more than thirty-five years, she has allowed and encouraged me to write from the heart: the most fantastic gift. Friends who have added immeasurable joy and provided invaluable camaraderie for at least forty years—and, in the last case, more than seventy—are precious beyond imagining, each in his or her way as heroic as amiable: George Gibson, Sanford Schwartz, Mickey Cartin, Ruth Agoos Villalovos, Kathy Agoos, Ted Agoos, Peter Agoos, Julie Agoos, Hallie Thorne, and Prudence Elizabeth Glass. This book is about the test of time, and they have aced it. Jane Safer, whom I have known from my earliest beginnings, has a warmth and intelligence that have never failed.

If the Alberses' marriage has shown the way that the deepest love and respect and shared values can survive copious challenges, the remarkable Kathy—my wife, Katharine Weber—has taken patience and courage to the extreme. With this book in particular, her gift for language and unrivaled capacities as a writer have been a boon. Our love, and the extraordinary connection we have, sustains me.

I thank my wonderful family for their insightfulness and their generous advices, and above all for the feeling of love. Nancy Weber has, as always, been both supportive and giving, especially with her guidance toward *Moby-Dick*. Robbie Smith has been his usual bright, energetic, forthright self: a boon to his father-in-law's life. Charles Lemonides, Ellen Weber Libby, Daphne Astor, and Micky Astor exemplify the warmth, wit, and connectedness of the relatives one chooses—and cherishes.

My daughters have been dreams while I have worked on this book—although they are always fantastic anyway. I know better than to put them in the same sentence. Lucy Swift Weber: you are the trouper to beat them all, beautiful beyond words for your warmth, your perceptiveness, and your generosity of heart. Charlotte Fox Weber: is there anyone on earth more insightful, more able to connect, and more accomplished in every way? (Charlotte is right up there with Le Corbusier and Steve Jobs as proof that the hidebound teachers stuck in limited minds should reconsider their notions of what truly smart people with gumption can do while the ones who only check the boxes of rote learning lag behind.) My God, how I love you two beautiful human beings!

And then there is Wilder Fox Smith; he is not even five years old as I write this. Yet he is already *everything:* unbelievably funny, kind from his inner depths, imaginative beyond all ken, and making leaps of mind and body at a rate and with capacities that seem, to his grandfather and pal and, I think, soul mate, unprecedented in human history. (I know, grandparents are impossible, but you have to meet this boy!)

I get teased for long-winded acknowledgments. I will keep the rest of these Bauhaus and iPhone succinct. But my appreciation of my generous friends is infinite. About two years ago, over one of our wonderful catch-up lunches at the Metropolitan Museum of Art, I mentioned to John Eastman that I considered the iPhone the realization of the Bauhaus dream. He immediately replied, "Nick, you should write a book about that." I thank him for the impetus to do so; it shows *his* out-of-the-box thinking. I am grateful in different ways, for each, Sean O'Riordain, Danjoe O'Sullivan, Regina Tierney, Anne Sisco, David Lieber, Samuel Gaube, Emma Lewis, David Zwirner, Pierre-Alexis Dumas, Lucas Zwirner, Lottie Hadley, Romain Langois, Sophie Dumas, Shane O'Neill, Seamus O'Reilly, Mike Adler, Maurice Moore, Fiona Kearney, Robert Devereux, André Tamone, Ray Nolan, Margaret Jay, Alan Riding, Brenda Danilowitz, Peter Deasy, Matthias Persson, Hans Renders, Hugh O'Donnell, Jaime Barry, Allegra Itsoga, Massamba Camara, Moussa Sene, Magueye Ba, Stefan Stein, Monica Zwirner, Victor Teheiu, Katinka Weber, Ramazan Cansu, Mark Sommerfeld, Alex Guillaumin, Mareta Doyle, Conor Doyle, Tom Doyle, Toshiko Mori, Gloria Loomis, Adhiraj Shakhawat, Francois Gibault,

Acknowledgments

Dario Jucker, Mary O'Reilly, Seth Cameron, Shannon Hart, James McAuley, Ross Ferrara, Fabrice Hergott, and Sam Haverstick.

For help specifically for this book, I can only say that it would not have been possible without the brilliant and inexhaustible Willem Van Roij. Philippe Corfa's patience and sterling intelligence have been amazing. Marc Jaffee has been diligent, tireless, and infallibly kind. One has an odd relationship with a copy editor: a person one has never met personally. In this case, Patrick Dillon, although I cannot even picture him, and know little about him, has pored through these pages with an attentiveness and an intelligence that makes me like him immensely. The splendid Lisa Sornberger has been an exemplar of grace and fortitude, and an emotional mainstay. Édouard Detaille has been saintly in his patience and off the charts in his sheer kindness and phenomenal insightfulness. Meir Kryger, a brilliant sleep doctor whose book on representations of sleep in art exemplifies the marriage of steely knowledge with poetic reach, spontaneously invented the title *iBauhaus* after asking me what I was writing at the time I started this project; I will never forget the warm smile with which he proffered what I consider to be his stroke of genius. And to the exceptional William Clark, bravo and deepest thanks; you embody the values and patience and tenacity that can make a literary agent the greatest of allies.

Notes

vii Don't think that our most: *The Letters of Paul Cézanne*, edited and translated by Alex Danchev (Los Angeles: Getty Publications, 2013), p. 358; translation slightly reworked by NFW.

Part II

66 Ironically, in the mid-1980s: Isaacson, Walter, *Steve Jobs* (New York: Simon & Schuster, 2011), p. 242.

66 "Steve believed it was": ibid., pp. 242–43.

Part III

69 Jobs's "great name": Kahney, Leander, *Jony Ive: The Genius Behind Apple's Greatest Products* (New York: Portfolio/Penguin, 2013), pp. 128–29.

70 A product manager: Segall, Ken, *Insanely Simple* (New York: Portfolio/Penguin, 2012), p. 105.

70 This was not the most literate: ibid., p. 207.

70 "From some companies": ibid.

Part IV

89 "When you open the box": Wingler, Hans, *The Bauhaus* (Cambridge, MA: MIT Press, 1972), p. 72.

90 In the conviction that: ibid., p. 109.

91 Gropius characterized: ibid.

91 "The creation of standard": ibid., p. 110.

91 The requisites were: ibid.

91 They would be manufactured: ibid.

92 "The attempt to penetrate": ibid., pp. 113–14.

92 The beauty of industrial objects: ibid., p. 114.

93 "We're talking about perception": ibid., p. 220.

94 The idea of an iPhone: ibid., p. 218.

94 The words that Ive used: ibid., p. 221.

Part V

106 Wallace Stevens, who lived in Hartford: Stevens, Wallace, *Opus Posthumous* (London: Faber & Faber, 1959), p. 294.

106 Stevens was like both: ibid., p. 296.

107 Stevens writes of people worldwide: ibid.

Part VI

129 On the day when the iPhone: Isaacson, *Steve Jobs*, p. 435.

131 "Eichler did a great thing": Isaacson, Walter, "How Steve Jobs' Love of Simplicity Fueled a Design Revolution," *Smithsonian*, September 2012.

131 "I love it when": Isaacson, *Steve Jobs*, p. 7.

Part VII

138 Ralph Tabberer—a colleague: Kahney, p. 3.

138 If they were walking: ibid., p. 4.

139 A drummer in a rock band: ibid., p. 20.

142 "If you can do without": Hanness Beckmann.

144 Ive also adhered to: Kahney, p. 22.

151 What had gone wrong at Apple: ibid., p. 101.

151 But then Steve Jobs said that: ibid., pp. 101–02.

Part IX

166 "They came close to": Isaacson, *Steve Jobs*, p. 11.

Part XI

186 Schapiro then quotes: Schapiro, Meyer, "*The Apples of Cézanne: An Essay on the Meaning of Still-life*" (New York: *Art News Annual*, 1968), p. 8.

Part XII

202 People who marched to: Patrick Hanlon, "Meet the Man Who Saved Apple," *Forbes,* December 6, 2015.

203 He emerged from the bar: Thomas J. Watson Jr. and Peter Petre, *Father & Son Co.: My Life at IBM and Beyond* (New York: Bantam Books, 1991).

203 In the same period: Frederick E. Allen, "Hitler and IBM," *American Heritage*, vol. 52, no. 5 (July/August 2001).

205 Tanimoto's response was: *Forbes,* 2015.

209 All quotations on pp. 209 through 212 in this section are from Foulkes, Nicholas, "Lunch with the FT: Jony Ive on the Apple Watch and Big Tech's Responsibilities," *Financial Times,* October 19, 2018.

Index

Page numbers in *italics* refer to illustrations.

Aachen, 45, 79
Aalto, Alvar, 119
abstract expressionism, 48
abstraction, 32, 78, 92, 99, 136, 220
Academy of Athens, 228
accident, 235–6
Adam and Eve, 177, *177*, 178, *178*, 179, 181, 192, 197
adaptation, 106
advertising, 69, 89; Apple, 69, 89–90, 116, 173–4, 185, 199–205; Intel, 89; "Think Different" campaign, 28–9, 199–203, *204*, *205*
Africa, 135, 154
African Americans, 133
airplanes, 124; Junkers, 216–17
Aix-en-Provence, 163
Albers, Anni, 10–12, 13, 15–16, 42–3, *43*, 44–62, 74, 94, 100–101, 201–2, 211, 232; background of, 45, 46, 46, 55–6, 60–1, 107, 133–4, 212, 226; Bauhaus ideals and, 42–62, 105–13, 121, 139, 212, 215–18, 226, 235; Connecticut home of, 16, 43, 45, 50–4, 60, 85–6, 105–9, *109*, 110–12, 153, 208, 225; death of, 41, 59; gravestone of, 59; interest in printing, 57–60; Judaism of, 133–4; Klee and, 215–18; marriage to Josef, 56, 60, 71, 133–4; meets Josef, 55; name change, 71; telephone and, 60–1; textiles by, 48–9, 57, *57*, 78, 94, 99–100, 108, 111, 143–4, 169, 201, 222, 229; Tupperware and, 56, *56*
Albers Josef, 10–13, 40, 42–3, *43*, 44–62, 94, 100–101, 161, 201–2, 211, 232; alphabet by, 208–9; background of, 45, 56, 107, 133–4, 169; Bauhaus ideals and, 42–62, 105–12, 118–19, 121, 140–4, *144*, 155, 164–5, 209, 218–19, 225; at Black Mountain College, 136–7, 140; CBS special on, 52–3; Cézanne and, 163–5; collage of Herbert Bayer, 15, *15*, 16; color and, 53, 93–4, 116–17, 127, 155–6, 218–19, 221, *221*, 227; Connecticut home of, 16, 43, 45, 50–4, 60, 85–6, 105–9, *109*, 110–12, 153, 155, 208, 225; Corning Glass murals, 126–7; death of, 16, 41, 59, 221; fame and success of, 46–7; glass constructions, 222; gravestone of, 59; *Homage to the Square* series, 45–6, 52, 58, 116–17, 155–6, 218–19, 221, *221*, 223; *Machine Art* exhibition catalogue cover, 129; marriage to Anni, 56, 60, 71, 133–4; meets Anni, 55; name change, 70–1;

Albers Josef (*continued*)
  photocopy machines and, 47–8;
  poetry by, 191–2; racial attitudes,
  133–4; as a teacher, 136–7, 140–4,
  144; teacup design, 144–6, *146*; on
  Vierzehnheiligen, 118, 119, *119*;
  workshop of, 155–6
Albers, Lorenz, 107, 133, 169
Albers Foundation, 10, 11–12, 211
Alessi, 149
Alexander the Great, 157
All One Farm, 87, 187, 194
aluminum, 128, 141, 18/
Amelio, Gilbert, 150
anima mundi, 220
anti-Semitism, 17, 75, 133–4, 203, 205
Apollo spacecraft, 24
Apple Computer, 5, 9, 19, 21–8, 36, 68,
  149–52, 172, 173–81; advertising, 69,
  89–90, 116, 173–4, 185, 199–205;
  Apple I, 81–3; Apple II, 24, 89,
  173; Apple III, 24, 25, 26, 26, 27;
  Bauhaus ideals and, 33, 35–41,
  82–98, 99–108, 111–25, 129, 137–43,
  154–7, 165–76, 184, 206, 222–36;
  beginnings of, 81–3, 86–90, 173–4;
  clothing, 91, 144, *144*, 170, *171*;
  computers, 19–30, 68–70, 81–90,
  150–2, 172; Cupertino headquarters,
  69, 92, 150; design failures, 24–8, 39,
  employees, 30, 91, 92; iMac, 68–70,
  83, 88; influence of Cuisinart on,
  82–5; iPhone design, 93–6, *96*, 97–8,
  101, 104, 126, 128, 137–43, 160,
  206, 225; irritation with, 157–61; Ive
  becomes designer for, 150–2; Jobs
  forced out of, 79, 149; Jobs returns
  to, 79, 150–1, 199, 201, 204; Lisa,
  24, 26–7, 27, 28–9, 58; LisaDraw,
  24; "lockdown" mode, 93; logo, 32,
  37, 39, 68, 89–90, 96, *96*, 159–60,
  173–5, *175*, 176, 178, 181, 184–97;
  marketing, 89–90, 123, 159–60,
  173–4, 199–205; materials and
  surfaces, 36–7, 72, 81–2, 96, 104,
  108, 125–9, 141, 170–1; name,
  68–70, 77–8, 87–9; naming of
  "i" devices, 68–70, 72, 77–8;
  Newton MessagePad 110, 150;

"1984" commercial, 201–2, 204;
  packaging, 37, 123; profits, 92;
  Project Juggernaut, 149–50; "Purple
  Dorm," 93; screens, 116–17, 120–2,
  122; "Snow White" concept, 150;
  stock ownership, 30, 92, 212; sued
  by Samsung, 96; "Think Different"
  campaign, 28–9, 199–203, *204*, 205;
  vocabulary, 22–4, 28–30, 94, 116. *See
  also specific products*
Apple I, 81–3
Apple II, 24, 89, 173
Apple III, 24, 25, 26, 26, 27
apples, 87–8, 160, 173–9, 232; by
  Cézanne, 183–6, *186*, 187–91, *191*,
  192, *192*, 193–7; as forbidden fruit,
  177–9, 181, 192, 197; sex and,
  177–9, 181, 188–91
Apple Stores, 37
Apple Watch, 210
Aristotle, 29–30, 49, 94, 228, 234
Armetide, 149
art world, 10, 161
Asia, 70, 96, 135
Aspen Institute, 5–12, 19, 104; Jobs's
  talk at, 5, 8–10, 19, *19*, 20–4, 27,
  28–31, 58, 94, 152, 234
Association of Arts and Industries, 137
audio system, 149, 151, *151*
Auschwitz, 17, 205; gas chambers,
  17–18
Austria, 13
automobiles, 24, 25, 41, 54, 58, 84–5,
  110, 132, 134, 212; design flops,
  25–6; gear shift, 97; seats, 86
Aztecs, 51

Bach, Johann Sebastian, 68, 131
Bacon, Francis, 48
Baden-Baden, 45
Bad Staffelstein, 118
ballet, 76, 77, 120, 168, 223
Barcelona chair, 170, 231
Barr, Alfred, 14, 136
Bauhaus, 3–4; aesthetics and values,
  3–4, 11, 15, 23, 26, 29–42, 48, 61,
  72, 82, 84, 87, 99–108, 112–13,
  153–9, 165–76, 197, 200–201, 206–9,
  212–13, 215–36; Alberses and,

42–62, 105–13, 121, 140–4, 164–5, 209, 212, 215–19, 235; attitude, 119, 164, 218; Bayer designs, 6–7, 7, 8, 13–15, 17, 80, 136, 222, 231; centenary, 208; closing of, 12, 13, 16, 45, 136, 229; clothing, 91, 144, *144,* 170, 171; communal meals, 30; design failures, 24, 158–9; in Dessau, 4, *5,* 13, 40, 42, 80, 85, 87, 90–3, 99, 106, 109, 111, 113, *113,* 121, 122, 136, 144, *144,* 158, 167, 168, 172, 207, 215, 217, 220, 226, 230, 234; finances, 24, 92, 199; first exhibition in America, 136; foundation course, 140–4, *144,* 165; founding of, 3–4, 31, 66; Gropius as director of, 3–4, 20–4, 30–1, 42, 61, 62, 79, 80, 90–3, 100, 112–14, 179–80, 199, 207, 226, 234; international reverence for, 135–37; iPhone and, 33, 35–41, 89, 92–8, 99–108, 111–25, 129, 137–43, 154–7, 165–76, 184, 206, 222–36; irritation with, 157–61; Jobs's reverence for, 8, 15, 20, 29–33, 66, 82–98, 125–6, 129, 131–2, 141, 151–2, 166–74, 184; laboratories, 87, 91, 234; manifesto, 61, *61,* 62; Masters, 82; materials and surfaces, 72, 107–8, 117, 140–5, 165, 170–2, 219; name, *5,* 67–9, 70, 77, 78–80; Nazism and, 12, 13, 16–18, 45; 1923 Weimar exhibition, 6–7, *7,* 8–9, 124, 136, 199, 226; 1938 MoMA exhibition, 14–15, 17; Plato and, 30, 31, 42, 80, 222–36; prototypes, 91, 100; racial prejudice, 133–4; terminology, 67–9; Wassily armchair, 39–40, *40,* 159, 225; in Weimar, 4, 6–8, 31, 40, 42, 57, 63, 67, 87, 91, 99, 109, 121, 135–6, 168, 171, 207, 210, 215, 234; whiteness, 111–22; "work for hire" system, 92; workshops, 87, 165, 172, 234
Bauhaus (journal), 91
Bauhaus: 1919–1928 (1938, Museum of Modern Art), 14–15, 17
"Bauhaus style," 44
Bayer, Herbert, 6–8, 10, *11,* 12–14, *14,* 15, *15,* 17, 26, 42; affair with Ise

Gropius, 16; Albers's collage of, 15, *15,* 16; Aspen Institute and, 6, 8, 11–14, 31, 104; Bauhaus design, 6–7, 7, 8, 13–15, 17, 80, 136, 222, 231; Nazi affiliations, 14, 16–18, 19, 203; paintings by, 16–17; personal allure of, 15–16; posters and catalogues, 6–7, *7,* 17, 18, 136, 222; Wardrobe for a Gentleman, 80
beauty, 224, 236
Beckmann, Hannes, 140–1
Beckmann, Max, 54
beehives, 50
Beethoven, Ludwig van, 100
Benefactor magazine, 74
Beowulf, 71
Berckelaers, Fernand-Louis, 73
Berlin, 4, 13, 24, 46, 49, 56, 62, 107, 167
Bible, 71
bicycles, 85
Biedermeier chairs, 121
birds' nests, 50
Blackberry, 36, 98
Black Mountain College, 17, 136–7, 140
Blade Runner (movie), 201
Blair, Tony, 138
Bloomingdale's, 44
blue, 114
Bluetooth, 102
BMW, 85
book design, 59
Bottrop, 56, 107
Brancacci Chapel, 178, 179
Brancusi, Constantin, 32
branding, 88
Brandt, Marianne, 40, 44, 217, 223; table lamp, 40, 217
Braun, 148, 149, 151, *151*
Braun, Max, 149
Brennan, Chrisann, 28
Brennan-Jobs, Lisa, 27–8, 66, 172, 181
Breuer, Marcel, 10, 39, 42, 85, 106, 168, 226; Jewishness of, 134; playpen, 158–9; Wassily armchair, 39–40, *40,* 159, 225
brick, 130
Browning, Robert, "Andrea del Sarto," 226

Brunner, Robert, 149, 150
Buchenwald, 17
Buddhism, 131–2, 201

Calcutta, 135–6
California, 30, 32, 69, 92, 101, 130,
    147, 148, 166; modernist houses,
    130–2, *132*, 133–4
cameras, 101; iPhone, 40–1, 124; Leica,
    54; Polaroid SX-70, 51, 54, *55*
capitalism, 193
carpentry, 91, 110, 146, 165
Cartier-Bresson, Henri, 43, 54, 110,
    219
Catholicism, 56, 118
CBS, 52
Cendrars, Blaise, 73
ceramics, 117
Cézanne, Paul, 101, 163–7, 183–94;
    apples by, 183–6, *186*, 187–91, *191*,
    192, *192*, 193–7; *Judgment of Paris*,
    186, *186*; late still-life, 192, *192*; self-
    portrait, 185, *185*
chairs, 3, 93, 121; Barcelona, 170,
    231; Bauhaus failures, 24; Dessau,
    93, 124; Memphis armchair, 149;
    Wassily armchair, 39–40, *40*, 159,
    225
Chanel, Coco, 120
Chartres cathedral, 49, 61
Chemcor, 127–8
Chemex coffeemaker, 41, 56
Chiat\Day, 69, 202, 205
Chicago, 5, 60, 137
children, 124; Bauhaus cradle, 124–5,
    158, *158*; Breuer playpen, 158–9;
    iPhone and, 101; puppets and,
    230
China, 150
Christopher, Saint, 216
chrome, 72
Churchill, Winston, 226
"c-I," 69–70
circuit boards, 82, 86
cleanliness, 120
clothing, 75, 80, 91; Apple, 91, 96, *96*,
    150, 170–1; Bauhaus, 91, 144, *144*,
    170, 171; white, 112, 119–20
Clow, Lee, 205

color, 47, 91, 94, 114, 143, 158; Josef
    Albers and, 53, 93–4, 116–17,
    127, 155–6, 218–19, 221, *221*, 227;
    ancient Greek use of, 117; Apple
    logo, 173–5, *175*, 178; Cézanne
    and, 164; hue, 114; iPhone, 36, 37,
    112; Steve Jobs and, 53; Kandinsky
    on, 43; primary, 115; television,
    53; temperature, 155–6; whiteness,
    111–22
Columbia College, 195
Compaq, 152
computers, 19, 41–2, 101, 152, 159,
    201, 227; aesthetics of, 19–20, 29,
    152; Apple, 19–30, 68–70, 81–90,
    150–2, 172; e-mail mishaps, 161;
    first, 29, 124; history of, 21–30;
    screens, 116–17, 120–2
concentration camps, 17–18, 205
Connecticut, 10, 11, 16, 57, 106, 147;
    Albers house in, 16, 43, 45, 50–4,
    60, 85–6, 105–9, *109*, 110–12, 153,
    155, 208, 225; Plank House, 108–9,
    110–11
Conran, Terence, 139
Container Corporation of America, 14
control, 231–2
Cook, Tim, 213
Corning Glass, 126–9; Chemcor,
    127–8; Gorilla Glass, 126–9, 174,
    235
courage, 226–7
cradle, Bauhaus, 124–5, 158, *158*
Cranach the Elder, Lucas, 177; Adam
    and Eve painted by, 177, *177*, 179
Crete, 117
Crocker, Betty, 107
Cuisinart, 82–3, *83*, 84–5
cummings, e. e., 72
Cupertino, 69, 92, 140, 150, 234

deception, 227
Dell, 152, 201
De Lucchi, Michele, 148, 149
Denver, 9
d'Erlanger, Baroness, 120
Dessau, 4; Bauhaus, 4, *5*, 13, 40, 42, 80,
    85, 87, 90–3, 99, 106, 109, 111, 113,
    113, 121, 122, 136, 144, *144*, 167,

168, 207, 215, 217, 220, 226, 230, 234; chairs, 93, 124
Diaghilev, Serge, 120
Dietman, Erik, 231
Diogenes, 157
drip-dry fabric, 108
DuBrul, Antonia Paepcke, 10–12
Dürer, Albrecht, 178; *Eve*, 177, *177*, 178
dyslexia, 137, 138, 166

ebony, 144, 145, 146, *146*, 155
Edison, Thomas, 202, 204
Edsel, 24, 25, *25*, 26; failure of, 25–6
Egypt, 48
Ehrlich, Franz, 17
Eichler, Joseph L., 130–4; modernist
 houses by, 130–2, *132*, 133–4
Eichmann, Adolf, 205
Einstein, Albert, 202, 204
Electrostar, 84
e-mail, 161; mishaps, 161
emotion, 48, 49–50, 94, 102, 153
entrepreneurs, 27–8; beginnings of
 Apple, 81–3, 86–90; products named
 after their children, 27–8
Epson, 201
Ertl, Fritz, 17–18
Esslinger, Hartmut, 150
etching, 58
Eugenie, Princess, 211
Europe, 20, 70, 135, 149, 172. *See also
 specific countries*
experimentation, 87, 109, 137, 143,
 156, 164, 167–9
"Extrudo," 95

FaceTime, 227
Fagus Factory, 62, 103, *103*, 104
Federal Trade Union Headquarters, 49
Feininger, Lyonel, 61, 217; print of
 Gothic cathedral, 61, *61*
Filene, Abraham Lincoln, 76–7
Filene's, 76–7
Financial Times, 208
flat roof, 106, 130
flatware, 3, 98
Fleishmann, Toni Ullstein, 46, 60–1,
 107, 134
flip phone, 95, *95*, 160

Folkwang, Hagen, 164
fonts, 58
food, 100, 101; apple, 173–9;
 microwaved, 124; plastic containers,
 56, 56, 100; preparation, 124, 125;
 salad bar, 108
food processor, 82–3, *83*, 84–5
Ford, Edsel, 27
Ford, Henry, 27
Ford Motor Company, 25, 26
form, 91, 165; Platonic ideal,
 222–36
Formica, 86
Fortune, 13
Foulkes, Nicholas, 210
Fox Press, 57–60
France, 45, 83, 133, 163, 205, 211
Frankfurt, 148, 149, 151
Freud, Sigmund, 143, 194
Fritz (cat), 43
Frost, Robert, "The Road Not Taken,"
 226
Fuller, Buckminster, 153
functionalism, 23, 41, 48, 54, 80, 91,
 92, 100, 131, 143, 154, 212
furniture, 3, 24, 39–40, 86, 93, 124,
 111, 134, 149, 158–9, 170, 225. *See
 also specific furniture*

Gandhi, Mahatma, 202, 204
Gauguin, Paul, 186–7
Gehry, Anita, 76
Gehry, Frank, 75–6; name change,
 75–6
Geldzahler, Henry, 110
General Electric, 14
geometry, 74, 158, 159
Germany, 4, 17, 36, 56, 60–1, 104, 140,
 148, 151, 163–4; Braun designs, 148,
 149, 151, *151*; Nazi, 12–14, 16–18,
 45, 61, 203–5; post–World War I
 inflation, 210; rococo churches,
 117–18, *119*; society, 167, 172; Third
 Reich, 12, 13, 17–18, 45, 197; World
 War I, 62–3; World War II, 149, 205.
 See also Bauhaus
"German Youth in a Changing World"
 (brochure), 18
Giacometti, Alberto, 196

Index

glass, 72, 102, 103, *103*, 108, 144, 145, 165, 170; Gorilla, 126–9, 174, 235; iPhone screen, 125–9, 160, 174

glassware, 85, 144–5, 146, *146*

Glueck, Grace, 52

Goertz, Albrecht, 84–5

Goethe, Johann Wolfgang von, 6, 8, 63

Goldberger, Paul, 76

Google, 36

Gorilla glass, 126–9, 174, 235

Gothic architecture, 49, 61

Graham, Martha, 153

Grand-Ducal Saxon Academy of Art, 67

Grand-Ducal School of Arts and Crafts, 67

graphic design, 6–7, 7, 29, 58, 59, 140, 165, 173; Apple logo, 32, 37, 39, 68, 89–90, 96, *96*, 159–60, 173–5, *175*, 176, 178, 181, 184–97; by Herbert Bayer, 6–7, 7, 8, 13–15, 17, 136, 222, 231; Lisa, 26–7, 29; by Jan Tschichold, 58, 59

gravity, 29, 118

Gray, Eileen, 115, 148, 149

Great Britain, 49, 97, 137–9, 146, 148, 174, 208–11; design education, 138–40

Greece, ancient, 29–31, 73–4, 104, 117, 156–7, 190, 222–36

Gropius, Alma Mahler, 42, 44, 64–6, 179–80

Gropius, Ise, 16

Gropius, Martin, 62

Gropius, Walter, 3–4, 6, 8, 20, 21, 22, 22, 35, 42, 121; Alma and, 64–6, 179–80; background of, 45, 62–5, 79; as Bauhaus director, 3–4, 20–4, 30–1, 42, 61, 62, 79, 80, 90–3, 100, 112–14, 179–80, 199, 207, 226, 234; Bauhaus manifesto, 61, *61*, 62; Dessau headquarters, 4, 111, 113, *113*; Fagus Factory, 62, 103, *103*, 104; founding of Bauhaus, 3–3, 20, 66; at Harvard, 137; Mies van der Rohe and, 78–9; 1923 Bauhaus exhibition and, 8–9, 124, 199, 226;

personal style of, 22–3, 62–6; private life of, 63–6, 179–80; tenacity of, 63

Gucci, 207

Hamburg, 74

Hamilton, William, 77

hardware stores, 50–1, 75

Hartford, Connecticut, 106

Harvard Coop, 136

Harvard Graduate School of Design, 137, 169

Harvard Society for Contemporary Art, 136

Harvard University, 114

Harvard University Art Museum, 10

Haus am Horn, 124

health care, 102

heart, 102

hemp, 72, 96, 143

Henley, W. E., "Invictus," 226

Hewlett-Packard, 24, 26

Himmler, Heinrich, 205

Hinduism, 120, 136

Hitachi, 201

Hitler, Adolf, 18, 203–5, 224

Hitler Youth, 18

Hochschule für Gestaltung, 140

Holland, 74, 115, 205

Hollerith, Herman, 205

"Hollerith cards," 205

Homebrew Computer Club, 86

Hoover vacuum cleaners, 78

Horace, 169

household appliances, 47, 51–3, 56, 78, 82, 90, 148–9; Cuisinart, 82–3, *83*, 84–5

hue, 114

human behavior, impact of iPhone on, *38*, 39–41, 206–7

IBM, 47, 201, 202, 203, 205

iMac, 68–70, 83, 88; name, 68–70

imagination, 231

immigrants, 75–7

Incas, 51

India, 131, 135–6, 235

India Society of Oriental Art, Fourteenth Annual Exhibition of, 135–6
Instagram, 161
Intel, 89
Internet, 69, 101, 206
iPad, 72, 77, 79, 89
iPhone, 32, 32, 33, 35–42, 60, 67, 72, 83, 89, 223; apple logo on, 37–9, 187, 192–3, 197; Bauhaus ideals and, 33, 35–41, 89, 92–8, 99–108, 111–25, 129, 137–43, 154–7, 165–76, 184, 206, 222–36; camera, 40–1, 124; case, 72, 125–9, 187, 206; colors, 36, 37, 112; construction, 36–9, 72, 95–7, 123, 125–9, 225; cost, 61, 207; design, 93–6, 96, 97–8, 101, 104, 126, 128, 137–43, 160, 206, 225; feel of, 96–7; flaws, 39; Gorilla Glass display, 126–9, 174, 235; human hand and, 36–7, 37; impact on human behavior, 38, 39, 40–1, 206–7; irritation with, 157–61; keyboard, 97, 98; launch of, 96, 96, 123, 129, 174, 206; malfunction, 39; marketing, 123, 159–60, 173–4; materials and surfaces, 36–7, 72, 96, 104, 108, 125–9, 141, 170–2, 187; medical uses of, 102; models, 223, 223; name, 68–9, 77–8; obsolescence, 206; packaging, 123, 159–60, 173–5; Platonic ideals and, 222–36; prototype, 126; sales, 206; screen, 95, 112, 116–17, 120–2, 122, 125–9, 160; technological advances, 104–5, 127, 222, 223; thinness and scale, 36–7, 37, 129, 142, 156, 168, 207, 222; whiteness, 111–22
iPod, 68, 72, 77, 79, 94, 106
Isaacson, Walter, 24, 27, 31, 131, 172
Israel, 45
Italy, 148, 178; Renaissance, 100, 120
Itten, Johannes, 21, 142–3, 171, 172
iTunes, 72
Ive, Heather, 148, 150
Ive, Jony, 93–7, 97, 126, 128, 137–52, 227, 229, 231, 232; background of, 97, 137–9; Bauhaus training of, 138–42; becomes designer for Apple, 150–2; dyslexia of, 137, 138; iPhone designed by, 93–8, 126; leaves Apple, 212–13; Newton MessagePad 110 design, 150; Orator design, 139, 147; Project Juggernaut design, 149–50; River Café interview, 208–11; wealth of, 212–13; Zebra pen design, 145–7
Ive, Mike, 97, 138, 139, 143, 144
Ive, Pamela Mary, 137–8

Jaguar XK 120, 41
Jandali, Abdul-fattah, 179, 180
Janis, Sidney, 52
Janoff, Rob, 173, 174
Japan, 20, 131, 142, 145, 147
Jeanneret, Charles-Édouard. See Le Corbusier
Jews, 56, 71, 75, 133; anti-Semitism and, 17, 75, 133–4, 203, 205; concentration camps, 17–18, 205; name changes, 71, 75–7
Job, Book of, 71
Jobs, Clara, 130, 166, 168, 180
Jobs, Patti, 130
Jobs, Paul, 71, 130, 166, 168, 170, 180
Jobs, Steven Paul, 5–7, 13, 16, 19, 19, 53, 58, 64, 71, 79, 96, 149–50, 173, 229, 232; aesthetics and values of, 20–1, 20–33, 47, 53, 65–6, 80, 81–98, 130–4, 150–2, 164–72, 193–4, 224, 232; Apple logo and, 173–9, 181, 184–97; Aspen Institute talk, 5, 8–10, 19, 19, 20–4, 27, 28–31, 58, 94, 152, 234; background of, 81, 130–2, 166–8, 170, 172, 179–81, 227; Bauhaus ideals and, 8, 15, 20, 29–33, 66, 82–98, 125–6, 129, 131–2, 141, 151–2, 166–74, 184; beginnings of Apple, 81–3, 86–90, 173–4; Buddhism and, 131–2; charisma of, 20; childhood home of, 130–2; color and, 53, 121; Cuisinart and, 82–3, 83, 84–5; death of, 197, 212–13; design failures, 24–8, 39; dyslexia of, 137, 166; education of, 166–7, 172; as a father, 27–8, 66, 181; forced out of Apple, 79, 149;

Jobs, Steven Paul *(continued)*
  glass screen and, 125–9, 174; iMac
  and, 68–70, 83, 88; iPhone design
  and, 93–6, 96, 97–8, 101, 104, 126,
  128, 137–43, 225; launch of iPhone,
  96, 96, 123, 129, 174; name, 71–2;
  naming of "i" devices, 68–70, 72,
  77–8; personal life of, 27–8, 64, 66,
  71, 179–81; personal style of, 21–3,
  82, 96, 96, 150–1, 170–1; at Pixar,
  150; return to Apple, 79, 150–1,
  199, 201, 204; tenacity of, 63, 128;
  "Think Different" campaign, 28–9,
  199–203, 204, 205; utopian ideas, 31;
  vocabulary of, 22–4, 28–30, 68–70,
  87–9; whiteness and, 121–2
Johns, Jasper, 47
Johnson, Philip, 129, 136
Jung, Carl, 220
Junkers aircraft, 216–17
jute, 143

Kahney, Leander, 140
Kandinsky, Nina, 42–3, 45, 89, 210
Kandinsky, Wassily, 7, 16, 31, 42–3, 45,
  78, 88–9, 99, 113, 122, 136, 158, 168,
  171, 176, 201, 210, 217, 220, 222,
  223, 226, 230; anti-Semitism of, 134;
  on Bauhaus faculty, 88; on color
  and sound, 43; sets for *Pictures at an
  Exhibition,* 168
Karsh, Yousuf, 54
Keler, Peter, 158; Bauhaus cradle,
  124–5, 158, 158
Kennedy, John F., 24, 226
keyboard, 81
King, Martin Luther, 236
King's College Chapel, 49
Kirstein, Lincoln, 76–7, 136
kitchen, 125
Klee, Felix, 43, 44, 230
Klee, Lily, 43–4, 45, 171, 221
Klee, Paul, 7, 16, 31, 42–5, 78, 88, 99,
  113, 122, 165, 169, 171, 176, 191,
  201, 210, 212, 215–23, 225, 226;
  Bauhaus years, 215–18, 220–2;
  puppets by, 229, 230
Kleenex, 78

Knoll, 201
Knoll, Florence, 201
Knossos, 117
Kokoschka, Oskar, 65–6
Kristallnacht, 205
Kyoto, 131

lamps, 3, 40, 149, 170; by Marianne
  Brandt, 40, 217
Land, Edwin, 55
Lauren, Ralph, 207
Le Corbusier, 49, 73, 115, 119, 235;
  name change, 73; Punjab Congress
  Assembly Hall, 235; Ronchamp
  church, 49; Unité d'habitation,
  133
Leica camera, 54
Leipzig, 167
Life, 13
lighting fixtures, 3, 40, 85, 149, 170
"Likeler" style, 131
Lincoln, Abraham, 76, 77
linoleum, 107
liquid retina, 116
Lisa, 24, 26–7, 27, 28–9, 58
LisaDraw, 24
lithography, 58
Loewy, Raymond, 41, 85
London, 137–8, 148, 211; River Café,
  208–10
loneliness, 41
Los Angeles, 69, 85
Louis, Morris, 55
LoveFrom, 213
lowercase vowels, 68–70, 72, 77

Mac computers, 83, 88, 93, 131
Machu Picchu, 50
Macintosh, 69
MacMan, 69, 70
Macworld Conference (2007), 96, 96
Macy's, 82, 83, 85
Mahler, Gustav, 64, 65, 66, 179
Mann, Thomas, 64
marble, 117, 155
marketing, 88–90; Apple, 89–90, 123,
  159–60, 173–74, 199–205
Marseille, 133

Masaccio, 179; *Expulsion from the Garden of Eden*, 178, *178, 179*
Masolino, 179
mass production, 36, 58, 91, 92, 128, 130
Mayan art, 51
McCarthy, Mary, 169
McKenna, Regis, 89–90, 173
meditation, 131
Melville, Herman, *Moby-Dick; or the Whale*, 114
Memphis armchair, 149
Memphis Group, 148
Mercedes-Benz, 54, 116
metalwork, 91, 146, 165
Mexico, 48, 54, 108
Meyer, Adolf, 103
Meyer, Hannes, 49
MGB GT, 58
Michelangelo, *David*, 117
mid-century modern, 111, 130–2, 132, 133–4
middle class, 132
Mies van der Rohe, 42, 44, 67, 100, 130, 155; background of, 45, 79, 169; Barcelona chair, 170, 231; Barcelona Pavilion, 130, 201; as director of Bauhaus, 67; Gropius and, 78–9; "Less is more," 155; racial attitudes, 134; Villa Tugendhat, 130, 222
Milan Duomo, 61
mindfulness, 123
minimalism, 155–6
MIT, 83
modernism, 3, 8, 70, 73, 76, 108, 110, 113, 121, 124, 130–4, 153
Moholy-Nagy, László, 137, 142–3
Mondriaan, Frits, 74
Mondrian, Piet, 57, 74, 114–15; name change, 74; whiteness and, 114–15
Morandi, Giorgio, 154
Morrison, Jasper, 148, 149
Motherwell, Robert, 55
Motorola StarTAC Rainbow, 95, *95*
mouse, computer, 121–2, 124
movies, 101, 201
Mozart, Wolfgang Amadeus, 103, 221
Muche, Georg, 91–2

Munich, 210
Museum of Modern Art, New York, 14, 100, 136, 208, 209; *Machine Art* exhibition, 129; 1938 Bauhaus exhibition, 14–15, 17
music, 68, 100, 111, 131, 135, 154, 221
Muslims, 120, 133, 179

Nabokov, Vladimir, 153
names, 67–80; Apple, 68–70, 77–8, 87–9; Bauhaus, 67–9, 70, 77, 78–80; changing, 70–80; of "i" devices, 68, 70, 72, 77–8; importance of, 88; iPhone, 68–9, 77–8
NASA, 83
National Association of Home Builders, 133
nature, 50, 102
Naugahyde, 109, 111
Nazism, 12, 14, 16–18, 61, 203–5; Bauhaus and, 12, 13, 16–18, 45; Herbert Bayer and, 14, 16–18, 19, 203; concentration camps, 17–18, 205
NCR, 203
necessity, 229–30
neoplasticism, 115
networking, 88
Neumann, Balthasar, 118, 119
New Bauhaus, 137
Newcastle Polytechnic, 139, 142
Newman, Arnold, 54, 110
Newman, Barnett, 118
Newton MessagePad 110, 150
New York, 60, 85
New York City Ballet, 77
New Yorker, The, 77
New York Times, 18, 52
NeXT, 150
Nishibori, Shin, 96
Nitze, Paul, 10
Nokia 7600, 95, *95*
Northern Renaissance painting, 195–6
nostalgia, 104

Obama, Barack, 80, 211
Obama, Michelle, 211
Old Testament, 71

Index

Olitski, Jules, 55
Olivetti Lettera 22 typewriter, 41
Orator, 139, 147
Ortega y Gasset, José, 6
Orwell, George, *1984,* 201, 204
Osthaus, Karl Ernst, 163–4

Paepcke, Elizabeth, 6, 10
Paepcke, Walter, 5–6, 9, 10–11, *11,*
    12, 14, 16; Aspen Institute and,
    10–12, 14
Palm Springs, 131
Panofsky, Erwin, 196
papacy, 120
paper folding, 140–1, 142, 144, *144*
Paragon, 85
Paris, 45, 57, 74, 115, 189
Parthenon, 117, 156
Penguin Books, 59
Philips, 149
Philostratus, 190
photocopy machines, 47–8
photography, 165; iPhone, 40–1, 124
Picasso, Pablo, 60, 74, 100
Piscator, Erwin, 24
Pitney Bowes, 139, 147
Pixar, 150
pixels, 58, 116
Plank House, 108–9, 110–11
plaster, 117
plastic, 82, 106, 108, 143; Cuisinart, 82,
    83, *83;* food containers, 56, *56,* 100
Plato, 30, 220, 222–36; aphorisms,
    222–36; Bauhaus and, 30, 31, 42, 80,
    222–36; *Laws,* 228; *Republic,* 228
playfulness, 229–30, 233
playpen, Bauhaus, 158–9
Plexiglas, 81
Plutarch, 157
plywood, 107
Polaroid SX-70 camera, 51, 54, *55*
polyester, 49
polyurethane, 108
Portland, Oregon, 87, 172
Powell, Albert, 116–17
practicality, 91, 92, 154
printing, 57–60
Project Juggernaut, 149–50
projector, 149; portable overhead, 139

Propertius, 187, 188–9
Protestantism, 56
punch-card technology, 203–5
purity, 56, 112, 120

racial discrimination, 133
Rams, Dieter, 41, 148, 149, 151; audio
    system, 149, 151, *151*
Rauschenberg, Robert, 53–4
rayon, 72
red, 114
Redse, Tina, 66
reductionism, 56, 155–6
redwood, 130
Reed College, 87, 112, 181, 227
Reich, Lilly, 44, 79
religion, 56, 71, 75, 118, 119, 120, 133,
    215–16
Rembrandt van Rijn, 74
Renaissance, 100, 120
Renaming Offices, 70
Renan, Ernest, 73
Renoir, Auguste, 194
"Reply All" function, 161
River Café, London, 208–10
Roberts Weaver Group, 139, 148
Robot-Coupe, 84
Rococo churches, 117, 118, 119,
    119
Rogers, Ruth, 211
Romanesque churches, 178
Rosenberg, Jakob, 196
Rouault, Georges, 163, 164
rubber, molded, 107
Rubinstein, Arthur, 6

Sabon, 58, 59, *59*
salad bar, 108
Salk, Jonas, 236
Salton yogurt maker, 51, 56
Samsung, 96; Apple sued by, 96
Samsung T700, 95, *95*
Sander, Jil, 74–5, 137
"Sandwich," 95
San Francisco, 147, 150, 180
sans serif typeface, 7, *7,* 13, 101
Sauser, Frédéric-Louis, 73
Schieble, Arthur, 180
Schieble, Joanne, 179, 180

Schlemmer, Oskar, 16, 42, 122, 223;
    Bauhaus Stairway, 233, 233; *The
    Triadic Ballet*, 168, 223
Schlumbohm, Peter, 56
Schoenberg, Arnold, 134
Schweitzer, Albert, 6
Scott, Ridley, 201, 204
screenprinting, 58
screens, 116–17; cracked, 129, 160;
    glass, 125–9, 160, 174; iPhone, 95,
    112, 116–17, 120–2, 122, 125–9, 160;
    liquid retina, 116
screen savers, 124
Sculley, John, 150
Sears, Roebuck, 56, 100
Segall, Ken, 69–70, 72, 199–200
Sert, Misia, 120
Seuphor, Michel, 73–4
Seurat, Georges, 57–8
sex, 151, 181, 194; apples and, 177–9,
    181, 188–91
Shapiro, Meyer, 184–96; on Cézanne's
    apples, 184–97; on van Eyck's *Rolin
    Madonna*, 195–6
Silicon Valley, 147–8
silk, 72, 96, 143
simplicity, 55, 91, 103, 154, 158
Singer sewing machines, 78, 203
smartphones, 33, 36, 41, 95, 95, 96,
    138, 160; cost of, 61–2; impact on
    human behavior, 38, 39, 40, 41,
    206–7; irritation with, 157–61;
    terminology, 95. *See also* iPhone;
    specific models and brands
Smith, David, 59, 60; *Standing
    Lithographer*, 59
Smith, Paul, 94–5, 211
snow, 119
Snowdon, Lord, 54, 110
Soby, James Thrall, 217
social class, 45, 60–2, 79, 107, 132,
    167–8, 212
social media, 125, 161
Socrates, 228, 229, 234
sofa, mid-century modern, 111
Sontheimer, Carl, 83–4
Sony products, 52, 53, 86, 95, 139
soul, 48–50, 220, 229; Bauhaus on, 48;
    Plato on, 220, 229

South America, 135, 140
Sparke, Penny, 140
speed, 123–4
spiders' webs, 50
Staatliches Bauhaus in Weimar, 67
stainless steel, 72, 124, 126, 127, 128,
    145
Stamford, Connecticut, 147
standardization, 91
Stanton, Frank, 52
Starmix, 84–5
steel, 72, 103, 103, 108, 114, 126, 127,
    128, 145
Stein, Gertrude, 72
Stella, Frank, 53
Stevens, Wallace, 106–7
Stravinsky, Igor, 136
Studebaker, 85
Supreme Court, U.S., 32
Switzerland, 45
synthetic materials, 49, 106, 107–9, 111
Syria, 180

Tabberer, Ralph, 138
tables, 3, 44, 51
tableware, 44, 85, 92, 97–8
tactile experience, 96, 143
Taiwan, 150
Talmud, 75
Tanimoto, Craig, 200–204, 229;
    "Think Different" campaign, 28–9,
    199–203, 204, 205
Tasso, *Aminta*, 190
tea glass and saucer, 144–5, 146, 146
telephone, 60, 139, 168; Anni Albers
    and, 60–1; Orator, 139, 147
television, 23–4, 52–3, 110, 124; Apple
    "1984" commercial, 201–2, 204;
    black-and-white, 53; Sony, 52, 86
Terramare Office, 18
textiles, 3, 45, 72, 91, 92, 107, 143, 159,
    170, 216, 226; by Anni Albers, 48–9,
    57, 57, 78, 94, 99–100, 108, 111,
    143–4, 169, 201, 222, 229
Theocritus, 190
"Think Different" campaign, 28–9,
    199–203, 204, 205
time sharing, 29
Torah, 76

Toronto, 75
toys, 3
"trial and error," 24
Trump, Donald, 133, 224, 232
Trump, Fred, 133
Tschichold, Jan, 58, 59
Tupper, Earl Silas, 56
Tupperware, 56, *56*
Turing, Alan, 174
Twombly, Cy, 47
TX2, 145
typefaces, 7, 58, 222; Sabon, 58, 59, *59;*
    sans serif, 7, *7*, 13, 101
typewriter, 41, 47, 98; Olivetti Lettera
    22, 41; stand, 86

Umbo, photograph of Bauhaus by, *144*
upholstery, 109, 111
"user-friendly," 36
utopianism, 31

van der Rohe, Georgia, 79
van Eyck, Jan, *Rolin Madonna,* 195–6
Vanity Fair, 97
velvet, 143
Venice, 39, 120
Verdon, Pierre, 84
video games, 29
Vienna, 17, 65
Vierzehnheiligen, 117, 118, *119*
vinyl, 111
Virgil, 187, 188–9
virginity, 120
Vlock, Laurel, 53
Vuitton, Louis, 207, 232

Walker, John, 136
Warburg, Edward, 136
Wardrobe for a Gentleman, 80
Wassily armchair, 39–40, *40,* 159, 225
Watson, Thomas John, Sr., 202–5
Wayne, Ron, 81
Weaver, Barry, 146–7
Weaver Company, 146–7

Weimar, 4, 6, 63, 70, 172; Bauhaus, 4,
    6–8, 31, 40, 42, 57, 63, 67, 87, 91, 99,
    109, 121, 135–6, 158, 168, 171, 172,
    207, 210, 215, 234; 1923 exhibition,
    6–7, *7,* 8–9, 124, 136, 199, 226
Werfel, Franz, 180
whiteness, 56, 82, 103, 111–22, 130;
    Albers house and, 105–6, *109,* 111;
    of Cuisinart, 82, 83, *83;* Mondrian
    and, 114–15
Wilder, Thornton, 6
Wilson, Woodrow, 203
window shades, 109, *109,* 111
Winsor & Newton, 219, 221
Wisconsin, 179, 180
Wolfe, Tom, *From Bauhaus to Our
    House,* 105
wood, 107, 108, 110, 117, 143
Woolf, Virginia, 153
World's Fair (1939, New York), 85
World War I, 9, 36, 62–3, 70, 226; end
    of, 135
World War II, 6, 11, 36, 84, 85, 149,
    180, 201, 205, 229
Wozniak, Steve, 86, 173
Wright, Frank Lloyd, 130, 132

Xerox, 78

Yale Graduate School, 57
Yale Law School, 78
Yale University School of Design, 140,
    169
yogurt maker, 51, 56
YouTube, 124

Zebra pen, 145–7
Zeiss, 139
Zen Buddhism, 131–2, 201
Zola, Émile, 187–8; *Mon Salon,* 188;
    L'Oeuvre, 188; *Le Ventre de Paris,*
    189–90
Zoroastrianism, 172
Zurich, 60

Illustration Credits

Page

5 Stiftung Bauhaus Dessau
(I 18983 F)

7 Digital Image © The Museum
of Modern Art / Licensed by
SCALA / Art Resource, NY /
© 2019 Artists Rights Society
(ARS), New York / VG Bild-
Kunst, Bonn

11 Photo by Gordon Coster / The
LIFE Picture Collection via
Getty Images / Getty Images

14 Keystone Press / Alamy Stock
Photo

15 © The Josef and Anni Albers
Foundation / Artists Rights
Society (ARS), New York, 2019

17 © Soichi Sunami,
photographer / Digital Image
© The Museum of Modern
Art / Licensed by SCALA /
Art Resource, NY

19 © 2019 Arthur Boden

22 Pictorial Press Ltd. / Alamy
Stock Photo

25 Pictorial Press Ltd. / Alamy
Stock Photo

26 INTERFOTO / Alamy Stock
Photo

27 INTERFOTO / Alamy Stock
Photo

32 dpa picture alliance / Alamy
Stock Photo

37 Anatoli Babii / Alamy Stock
Photo

38 (top) Antonio Guillem
Fernández / Alamy Stock
Photo

38 (bottom) David Kashakhi /
Alamy Stock Photo

40 Digital Image © The Museum
of Modern Art / Licensed by
SCALA / Art Resource, NY

43 © Henri Cartier-Bresson /
Magnum Photos

46 ullstein bild—Waldemar
Totzenthaler

55 Jeremy Pembrey / Alamy Stock
Photo

56 State Archives of Florida /
Florida Memory / Alamy
Stock Photo

A Note on the Type

The text of this book was set in Sabon, a typeface designed by Jan Tschichold (1902–1974), the well-known German typographer. Based loosely on the original designs by Claude Garamond (ca. 1480–1561), Sabon is unique in that it was explicitly designed for hot-metal composition on both the Monotype and Linotype machines as well as for filmsetting. Designed in 1966 in Frankfurt, Sabon was named for the famous Lyons punch cutter Jacques Sabon, who is thought to have brought some of Garamond's matrices to Frankfurt.

Composed by North Market Street Graphics,
Lancaster, Pennsylvania

Printed and bound by LSC Communications,
Crawfordsville, Indiana

Designed by Cassandra J. Pappas